Methods in Molecular Biology

Series Editor
John M. Walker
School of Life and Medical Sciences
University of Hertfordshire
Hatfield, Hertfordshire, UK

For further volumes:
http://www.springer.com/series/7651

For over 35 years, biological scientists have come to rely on the research protocols and methodologies in the critically acclaimed *Methods in Molecular Biology* series. The series was the first to introduce the step-by-step protocols approach that has become the standard in all biomedical protocol publishing. Each protocol is provided in readily-reproducible step-by-step fashion, opening with an introductory overview, a list of the materials and reagents needed to complete the experiment, and followed by a detailed procedure that is supported with a helpful notes section offering tips and tricks of the trade as well as troubleshooting advice. These hallmark features were introduced by series editor Dr. John Walker and constitute the key ingredient in each and every volume of the *Methods in Molecular Biology* series. Tested and trusted, comprehensive and reliable, all protocols from the series are indexed in PubMed.

Transgenic Mouse

Methods and Protocols

Edited by

Melissa A. Larson

Transgenic and Gene-Targeting Facility, University of Kansas Medical Center, Kansas City, KS, USA

Department of Molecular and Integrative Physiology, University of Kansas Medical Center, Kansas City, KS, USA

Humana Press

Editor
Melissa A. Larson
Transgenic and Gene-Targeting Facility
University of Kansas Medical Center
Kansas City, KS, USA

Department of Molecular and Integrative Physiology
University of Kansas Medical Center
Kansas City, KS, USA

ISSN 1064-3745 ISSN 1940-6029 (electronic)
Methods in Molecular Biology
ISBN 978-1-4939-9836-4 ISBN 978-1-4939-9837-1 (eBook)
https://doi.org/10.1007/978-1-4939-9837-1

Image courtesy of: Benoît Kanzler, Transgenic Mouse Core Facility, Max-Planck Institute of Immunobiology & Epigenetics, Freiburg, Germany

This Humana imprint is published by the registered company Springer Science+Business Media, LLC, part of Springer Nature.
The registered company address is: 233 Spring Street, New York, NY 10013, U.S.A.

Preface

Our collective ability to modify the genome of a mouse has existed for almost 40 years—a remarkable period of scientific achievements! During that time, transgenic technologies have allowed the generation of thousands of models for disease research and elucidation of signaling pathways and gene function. Far from these technologies remaining stagnant, the techniques and protocols employed to generate genetically modified mice continue to evolve at a rapid pace, necessitating the publication of timely resources. This volume will make every effort to provide both a historical foundation in standard protocols and a comprehensive update on the latest techniques used in making gene-modified mice for both novice and experienced personnel in the transgenic mouse field.

Kansas City, KS, USA *Melissa A. Larson*

Contents

Preface . . . . *v*
Contributors . . . . *ix*

1 The History of Transgenesis 1
Thomas L. Saunders

2 Pronuclear Microinjection of One-Cell Embryos 27
Melissa A. Larson

3 Integrase-Mediated Targeted Transgenics Through Pronuclear Microinjection 35
Ruby Yanru Chen-Tsai

4 In Vivo Validation of CRISPR Reagents in Preimplantation Mouse Embryos 47
Melissa A. Larson, Katelin A. Gibson, and Jay L. Vivian

5 Targeted Mutations in the Mouse via Embryonic Stem Cells 59
Marina Gertsenstein, Joffrey Mianné, Lydia Teboul and Lauryl M. J. Nutter

6 Blastocyst Microinjection with Embryonic Stem Cells 83
Melissa A. Larson

7 Generation of Large Fragment Knock-In Mouse Models by Microinjecting into 2-Cell Stage Embryos 89
Bin Gu, Marina Gertsenstein, and Eszter Posfai

8 Embryo Transfer Surgery 101
Melissa A. Larson

9 Nonsurgical Embryo Transfer Protocol for Use with the NSET™ Device 107
Barbara J. Stone

10 Nuclear Transfer and Cloning 113
Ling Liu

11 Sperm-Mediated Genetic Modifications 125
Marialuisa Lavitrano, Laura Farina, Maria Grazia Cerrito and Roberto Giovannoni

12 Genotyping Genetically Modified (GM) Mice 133
Neeraj K. Aryal and Jan Parker-Thornburg

13 Nomenclature: Naming Your Gene-Modified Mouse 149
Alicia Valenzuela

14 Breeding Strategies for Genetically Modified Mice 163
Jan Parker-Thornburg

15 Strategies for Behaviorally Phenotyping the Transgenic Mouse 171
Kenneth E. McCarson

16 Cryobanking and Recovery of Genetically Modified Mice 195
Toru Takeo and Naomi Nakagata

17 Simple Transportation of Genetically Engineered Mice via Cold Storage Techniques 211
Hidetaka Yoshimoto, Toru Takeo, and Naomi Nakagata

18 Reprogramming of Primary Human Cells to Induced Pluripotent Stem Cells Using Sendai Virus 217
Julia M. Draper and Jay L. Vivian

Index *235*

Contributors

NEERAJ K. ARYAL • *Genes and Development Program, Department of Genetics, Graduate School for Biomedical Sciences, MD Anderson Cancer Center, Houston, TX, USA; Bioscience, Oncology R&D, AstraZeneca, Boston, MA, USA*

MARIA GRAZIA CERRITO • *School of Medicine and Surgery, University of Milano-Bicocca, Monza, Italy*

RUBY YANRU CHEN-TSAI • *Applied StemCell, Inc., Milpitas, CA, USA*

JULIA M. DRAPER • *Transgenic and Gene-Targeting Institutional Facility, University of Kansas Medical Center, Kansas City, KS, USA*

LAURA FARINA • *School of Medicine and Surgery, University of Milano-Bicocca, Monza, Italy*

MARINA GERTSENSTEIN • *The Centre for Phenogenomics (TCP), Toronto, ON, Canada*

KATELIN A. GIBSON • *Transgenic and Gene-Targeting Institutional Facility, University of Kansas Medical Center, Kansas City, KS, USA*

ROBERTO GIOVANNONI • *School of Medicine and Surgery, University of Milano-Bicocca, Monza, Italy*

BIN GU • *Program in Developmental and Stem Cell Biology, Hospital for Sick Children, Toronto, ON, Canada*

MELISSA A. LARSON • *Transgenic and Gene-Targeting Facility, University of Kansas Medical Center, Kansas City, KS, USA; Department of Molecular and Integrative Physiology, University of Kansas Medical Center, Kansas City, KS, USA*

MARIALUISA LAVITRANO • *School of Medicine and Surgery, University of Milano-Bicocca, Monza, Italy*

LING LIU • *IVF Canada Fertility Center, Toronto, ON, Canada*

KENNETH E. MCCARSON • *Department of Pharmacology, Toxicology, and Therapeutics, Kansas Intellectual and Developmental Disabilities Research Center, University of Kansas Medical Center, Kansas City, KS, USA*

JOFFREY MIANNÉ • *The Mary Lyon Centre, MRC Harwell Institute, Oxon, UK*

NAOMI NAKAGATA • *Division of Reproductive Engineering, Center for Animal Resources and Development (CARD), Kumamoto University, Chuo-ku, Kumamoto-shi, Japan*

LAURYL M. J. NUTTER • *The Centre for Phenogenomics (TCP), Toronto, ON, Canada; Genetics and Genome Biology, The Hospital for Sick Children, Toronto, ON, Canada*

JAN PARKER-THORNBURG • *Department of Genetics, Genetically Engineered Mouse Facility, MD Anderson Cancer Center, Houston, TX, USA*

ESZTER POSFAI • *Program in Developmental and Stem Cell Biology, Hospital for Sick Children, Toronto, ON, Canada; Department of Molecular Biology, Princeton University, Princeton, NJ, USA*

THOMAS L. SAUNDERS • *Transgenic Animal Model Core, University of Michigan Medical School, Ann Arbor, MI, USA; Division of Genetic Medicine, Department of Internal Medicine, University of Michigan Medical School, Ann Arbor, MI, USA*

BARBARA J. STONE • *ParaTechs Corporation, Lexington, KY, USA*

TORU TAKEO • *Division of Reproductive Engineering, Center for Animal Resources and Development (CARD), Kumamoto University, Chuo-ku, Kumamoto-shi, Japan*

LYDIA TEBOUL • *The Mary Lyon Centre, MRC Harwell Institute, Oxon, UK*
ALICIA VALENZUELA • *The Jackson Laboratory, Bar Harbor, ME, USA*
JAY L. VIVIAN • *Transgenic and Gene-Targeting Institutional Facility, University of Kansas Medical Center, Kansas City, KS, USA; Department of Pathology and Laboratory Medicine, University of Kansas Medical Center, Kansas City, KS, USA*
HIDETAKA YOSHIMOTO • *Division of Reproductive Engineering, Center for Animal Resources and Development (CARD), Kumamoto University, Chuo-ku, Kumamoti-shi, Japan*

Chapter 1

The History of Transgenesis

Thomas L. Saunders

Abstract

A transgenic mouse carries within its genome an artificial DNA construct (transgene) that is deliberately introduced by an experimentalist. These animals are widely used to understand gene function and protein function. When addressing the history of transgenic mouse technology, it is apparent that a number of basic science research areas laid the groundwork for success. These include reproductive science, genetics and molecular biology, and micromanipulation and microscopy equipment. From reproductive physiology came applications on how to optimize mouse breeding, how to superovulate mice to produce zygotes for DNA microinjection or preimplantation embryos for combination with embryonic stem (ES) cells, and how to return zygotes and embryos to a pseudopregnant surrogate dam for gestation and birth. From developmental biology, it was learned how to micromanipulate embryos for morula aggregation and blastocyst microinjection and how to establish germline competent ES cells. From genetics came the foundational principles governing the inheritance of genes, the interactions of gene products, and an understanding of the phenotypic consequences of genetic mutations. From molecular biology came a panoply of tools and reagents that are used to clone DNA transgenes, to detect the presence of transgenes, to assess gene expression by measuring transcription, and to detect proteins in cells and tissues. Technical advances in light microscopes, micromanipulators, micropipette pullers, and ancillary equipment made it possible for experimentalists to insert thin glass needles into zygotes or embryos under controlled conditions to inject DNA solutions or ES cells. To fully discuss the breadth of contributions of these numerous scientific disciplines to a comprehensive history of transgenic science is beyond the scope of this work. Examples will be used to illustrate scientific developments central to the foundation of transgenic technology and that are in use today.

Key words Transgene, Transgenic mice, Superovulation, Transgenesis, Micromanipulation, Transgenic core facility

1 Introduction

The generation of genetically engineered mice to study gene function and for use as models of human disease can be seen as a continuation of the fascination of pet fanciers with mice displaying unusual phenotypes, such as the circling behavior of the Japanese waltzing mouse [1]. The interest of people in uncommon coat color patterns and behaviors led to collections of mice with odd banding patterns and behaviors. In time, scientists became

Melissa A. Larson (ed.), *Transgenic Mouse: Methods and Protocols*, Methods in Molecular Biology, vol. 2066,
https://doi.org/10.1007/978-1-4939-9837-1_1, © Springer Science+Business Media, LLC, part of Springer Nature 2020

interested in the study of pigmentation genetics research using mouse models. Clarence Cook Little was an early mouse geneticist whose early work includes the coat color genetics of mice. His work led to the founding of the Jackson Laboratory and recognition of the value of inbred mouse strains for transplantation research. The widely used C57BL and DBA came from his laboratory [2]. The history of the domestication of mice, cataloging of coat colors, and development of inbred strains are interesting topics in their own rights [3–5].

1.1 Advent of Molecular Biology

In the atomic age, numerous mouse mutants were generated in studies designed to quantify the effects of radiation on man [6]. Toxicology studies on chemical compounds to evaluate their safety showed that ethylnitrosourea (ENU) also introduced random genetic changes. In these experiments, the rationale was to subject mice to mutagenic treatments and then to evaluate the animals for observable phenotypes such as coat color changes. The description of the DNA double helix as the unit for heredity [7] was followed by molecular biologists who devised tools to experimentally manipulate DNA and to produce deliberate changes by "gene cloning" [8]. Today, cloning DNA and genes is a routine procedure. Complex genetic structures required to answer research questions can be rapidly assembled with rapidly advancing DNA synthesis technology [9, 10].

1.2 Reproductive Physiology

Basic science research in reproductive physiology and cell culture made important contributions to the practice of mouse transgenesis. The established methods of superovulation, cell culture, and need to synchronize manipulated embryos with pseudopregnant female recipient mice made it practical to generate transgenic mice after introducing nucleic acid solutions into zygotes by microinjection. The development of superovulation regimens evolved over time. The ability to culture zygotes in petri dishes in incubators also required significant optimization [11]. After the ovulatory effects of gonadotropins were observed, systematic studies on the PMSG and hCG doses, on timing of hormone administration, on the optimal age of superovulation, and on the effects of genetic background on superovulation were conducted to produce the standard treatments that are now widely employed in the field [12–14].

1.3 Advances in Equipment

The final element needed for microinjection of nucleic acids into zygotes or ES cells into blastocysts were the micromanipulators, microinjectors, and research microscopes that are under continual development by various companies. The development of micromanipulators followed the invention of microscopes. A variety of technologies were developed along the way, including sliding plate, thermal expansion, hydraulic, rack and pinion, and lever-activated mechanical micromanipulators [15]. The first mouse zygote microinjection study and the first transgenic mice were produced with the Leitz micromanipulator [15–17] (an updated version of this micromanipulator is available for purchase from Leica). Current micromanipulator products manufactured for use in mouse embryo micromanipulation use hydraulic (e.g., Narishige), lever-activated mechanical (e.g., Leica), or motorized technologies (e.g., Eppendorf and Sutter). Research microscopes have been developed that provide 400× magnification with differential interference contrast optics to enhance contrast in thick, low-contrast specimens such as mouse zygotes and embryos.

2 Superovulation

One of the basic requirements for successful mouse transgenesis is a reliable source of zygotes and embryos for microinjections. This is most often obtained by superovulation of juvenile mice with pregnant mare's serum gonadotropin (PMSG) and human chorionic gonadotropin (hCG).

2.1 Hormones

It was reported in 1928 that human pregnancy urine contained gonad-stimulating activity as detected by the treatment of 3- to 4-week-old mice [18]. The first demonstration of the gonad-stimulating activity of pregnant mare's serum in 18- to 23-day mice and rats was described by Cole and Hart [19]. Subsequent purifications of the gonadotropin from the serum allowed for refinements in PMSG treatments [20].

2.2 Timing and Dose

Studies on mouse superovulation explored the dose and timing of PMSG and hCG treatments to identify the time of ovulation and the optimal time to collect zygotes and embryos for experimental research. The standard timing for mouse superovulation is PMSG administration 60 h prior to ovulation and hCG administration 12 h prior to ovulation, as described by Runner and Gates [21]. They also reported that prepubertal female mice that received superovulation treatments did not become pregnant. The dose of PMSG and hCG was adjusted upward from initial doses of 1–2 IU [22] to 5 IU of PMSG and 5 IU of hCG [23]. Subsequent research established the regimen of 5 IU of PMSG and 5 IU of hCG administered 46–48 h apart before mating juvenile female mice with stud males [12, 24]. This is the current accepted best practice [14].

2.3 Yield

Mouse superovulation treatments can produce from 9 to 60 zygotes per egg donor depending on the age and genetic background of the animals [12, 13, 25–27]. Alternatives to PMSG and hCG have been tested. For example, the substitution of human menopausal gonadotropin for PMSG was found to result in superovulation but a failure to produce live pups or transgenic mouse founders [28]. The substitution of follicle-stimulating hormone for PMSG was found to significantly reduce the yield of transgenic rat founders [29, 30]. More recently, the use of anti-inhibin antiserum to superovulate mice has generated 100 oocytes from a single female mouse. The treatment can be used to produce oocytes for in vitro fertilization (IVF) procedures [31–33]. The zygotes produced by IVF can then be microinjected to produce genetically engineered mice.

3 Embryo Culture

Developmental biologists pursued methods to culture and observe mouse zygote development as they worked to elucidate steps in embryogenesis. Prior to the development of culture medium that supported the growth of preimplantation embryos, Runner [34] observed the development of these stages by introducing zygotes and late-stage embryos into the anterior chamber of the mouse eye. Intraocular growth permitted the observation of development through implantation. Such culture conditions are impractical for the purposes of transgenesis, which require frequent access to a pool of zygotes or embryos for micromanipulation.

3.1 Development of Culture Media

Synthetic culture media were developed over a period of years. Hammond [35] found that mouse morula develop into blastocysts after overnight culture in a synthetic medium (a saline extract of egg white) although two-cell embryos cultured in this medium did not progress. Whitten [36] reported that 0.1–0.4% bovine serum albumin (BSA) with glucose in Krebs-Ringer buffer supported the development of morula to blastocyst as effectively as the egg white extract. Whitten [37] then reported that the addition of lactate to the medium supported the development of two-cell embryos to the blastocyst stage. Further refinements in the medium resulted in Brinster's medium for ovum culture that permitted culture of random bred Swiss mice embryos from the two cell to blastocyst stage [38, 39]. A refinement of Whitten's medium (reduced NaCl concentration, use of 4 mg/ml BSA) and the use of genetically standardized mice showed that hybrid zygotes from (C57BL/10J X SJL/J)F1 could be cultured to the blastocyst stage, although zygotes from inbred C57BL/10J, inbred SJL/J, and inbred 129/Rr mice could not [40]. These authors also noted that zygotes from outbred mice failed to develop to the blastocyst stage.

3.2 Overcoming Two-Cell Block

Overcoming this phenomenon of two-cell block, by which it is meant that zygotes are blocked from further development in cell culture, was the next obstacle in developing a complete culture system for the mouse. CZB medium was found to support the

Table 1
Mouse zygote and embryo culture medium formulations

Medium name	Formulation described
Brinster's medium for ovum culture (BMOC)	Brinster and Biggers [39]
Whitten's medium (WM)	Whitten and Biggers [40]
M2	Quinn et al. [45]
M16	Whittingham [46]
CZB	Chatot et al. [41]
Potassium simplex optimized medium (KSOM)	Erbach et al. [43]
KSOM + amino acids	Biggers et al. [44]

Summary of mouse embryo media development over time. Currently, most laboratories use M2 to work with mouse embryos on the bench and KSOM + amino acids to culture mouse embryos in incubators

development of both outbred CF-1 zygotes and hybrid (C57BL/6J X SJL/J)F1 zygotes to the blastocyst stage with a wash from CZB to a glucose-supplemented CZB formulation [41]. Subsequently, it was demonstrated that zygotes from hybrid mouse strains that do not undergo two-cell block, when placed in CZB medium containing 5.5 mM glucose alone, developed to the blastocyst stage without the need to wash the embryos from one CZB formulation to another [42]. KSOM medium was then found to support the development of zygotes obtained from the mating of outbred CF1 females with (C57BL/NCrl X DBA/2NCrl)F1 males [43]. At the current time, many laboratories involved in mouse zygote and preimplantation embryo culture use the most recent formulation of KSOM that includes amino acids [44]. KSOM+AA supports robust and complete development of zygotes from CF1 X (C57BL/NCrl X DBA/2NCrl)F1 matings through to the blastocyst stage. Mouse embryo media discussed here are listed in Table 1. Reviews on the development of mouse embryo culture media are available for further reading [11, 47–49].

4 Embryo Transfer

The basic procedure of zygote and embryo transfer to the reproductive tract resulting in live born animals was first demonstrated in the rabbit [50] and then in the rat [51]. The mouse followed in 1937 [52]. Reviews of the progress of embryo transfer can be found in Runner and Palm [22] and in McLaren and Michie [53].

4.1 Embryo Numbers and Timing

The importance of synchronizing the gestational age of the embryos to be transferred with the pseudopregnant recipient reproductive tract was recognized early on in the mouse [53]. It was demonstrated that in transfers of mouse embryos to the uterine environment, the gestation period was the same as for natural pregnancies (19–20 days) [53]. Fekete and Little [54] also demonstrated that the typical numbers of embryos collected 2.5 days after copulation plug observation from naturally ovulated eggs were nine from DBA mice and eight from C57 black mice. These results were confirmed in experiments that tested the feasibility of transferring blastocysts. The number of eggs that can be collected from naturally ovulating mice is significantly lower than the number that can be collected from superovulated mice. Factors were identified that optimized the number of live born pups after uterine transfer, such as transferring embryonic day 3.5 (e3.5) blastocysts into recipients at e2.5 [53, 55]. Additionally, live born pup numbers increased with the observation that when ovaries do not show signs of recent ovulation (i.e., the presence of corpora lutea), pregnancies are unlikely to result after embryo transfer [53]. After surgical transfer of fertilized eggs to e0.5 pseudopregnant recipients, implantation begins 90 h after transfer, whereas morula begin to implant as early as 42 h post transfer [56].

4.2 Oviductal Transfer

The essential procedure of zygote transfer to the oviducts of pseudopregnant females was summarized in Gordon et al. [17] in the production of the first transgenic mouse that continues to be the standard procedure used today [14]. The majority of zygote transfers take the form of transferring microinjected pronuclear zygotes into both oviducts of pseudopregnant female mice. Asynchronous transfer of zygotes to the uterus results in implantation failure [57]. Refinements in oviduct transfer that have been tested include the comparison of the effectiveness of embryo transfer to the ostium or the isthmus of the oviduct [58] and refinements in surgical techniques to deliver embryos to the reproductive tract [59–61]. After the microinjection of zygotes with nucleic acids, the best results (as measured by birth rates) are obtained when 18 microinjected zygotes are divided between the two oviducts of a single recipient and transferred on the day of microinjection [62]. With respect to the introduction of embryos into a single uterine horn or oviduct, experimental evidence shows that embryos do not migrate from one uterine horn to the other uterine horn after embryo transfer. This has been tested repeatedly by different investigators over time and shown not to occur [63–65].

4.3 Uterine Transfer

Similarly, the surgical procedure for the transfer of blastocysts described by McLaren and Michie [53] is essentially the same procedure used today [14]. Blastocysts can be transferred either to the oviducts or to the uterus. Refinements in blastocyst transfer include the use of nonsurgical methods [66–69]. When the blastocyst delivery device is nonsurgically inserted into the pseudopregnant mouse, blastocysts are placed in either the right or the left horn by the device. However, the device does not allow the operator to choose into which uterine horn the blastocysts are placed or to place blastocysts in both horns; thus, half of the mouse's reproductive capacity is not used.

5 Genetics

5.1 Effect upon Superovulation

The response to superovulation is under the control of the genetic background of the mice in question. A survey of 16 inbred mouse strains showed that a range of responses to that varied from 9 eggs (A/J mice) to 53 eggs (C57BL/6J mice) per donor [13]. Other reports on the response of mouse strains to superovulation demonstrate that different strains of mice respond differently to superovulation because of genetic background and age [12, 13, 25–27, 70–79]. The use of anti-inhibin goat serum as a superovulatory agent has produced as many as 100 oocytes from a C57BL/6 mouse that can be used in IVF procedures [31]. In contrast to PMSG/hCG treatments that can be used to produce both pronuclear zygotes after natural mating and oocytes for IVF, the anti-inhibin procedure is only useful in the production of oocytes.

5.2 Effect upon Transgenic Efficiency

In addition to the effects of genetics on the response to superovulation, genetics affect the efficiency of transgenic mouse production. The microinjection of 1463 C57BL/6 eggs yielded 15 transgenic founders or one founder per 100 microinjected eggs, while the microinjection of 1201 (C57BL/6 X SJL)F2 resulted in 39 transgenic founders (three founders per 100 microinjected eggs) [80]. In a study that compiled data from tens of thousands of eggs, it was found that FVB/N microinjection produces three founders per 100 microinjected eggs, that (C57BL/6 X DBA/2)F2 and SW zygotes produced two, and that C57BL/6 produced one [75]. The largest study conducted to date, by the International Society for Transgenic Technologies, evaluated data from 1,135,446 microinjected zygotes from a large number of transgenic core facilities, and largely agreed with the earlier findings [62, 81]. This survey found that for every 100 microinjected (C57BL/6 X SJL)F2 zygotes, three founders were produced; 100 microinjected C57BL/6 embryos produced one founder, FVB/N produced one founder, and (C57BL/6 X CBA)F2 produced 1.5 founders (*see* **Note 1**).

6 Transgenes

6.1 Recombinant DNA Technology

The essential elements of a transgene are the promoter/enhancer, the sequence for the desired expression product (reporter protein, recombinase, regulatory RNA, etc.), and a polyadenylation signal for mRNA stability and protein expression. While it is impractical to provide an account of the history of recombinant DNA technology [82, 83], some important milestones can be noted: (a) development of enzymes that permit DNA fragments to be isolated and joined together in new configurations, (b) cloning of DNA fragments in bacteria and expression of proteins from cloned DNA in bacteria, (c) application of chemistry to the synthesis of DNA, such as oligonucleotides, and (d) invention of the polymerase chain reaction (PCR).

6.2 Application to Cells

Once recombinant DNA technologies were in hand, then it was a logical extension to test the ability to introduce recombinant DNA into model systems such as cultured cells or into model organisms such as the mouse. Initial experiments indicated that it was possible to express proteins from exogenous cloned DNA introduced into cultured cells by transfection [84–86]. These studies relied on drug selection to identify transformed cell clones that carried the genetic material and expressed herpes simplex virus (HSV) thymidine kinase. The calcium phosphate DNA precipitation procedure to transform cultured cells with DNA [87] described in these papers could not be applied to zygotes.

6.3 Transgenic Mice: Random Transgene Integration

6.3.1 Scientific Development

The production of transgenic mice by the microinjection of DNA molecules into the pronucleus of mouse zygotes required the convergence of numerous research fields, including mouse reproductive physiology (superovulation, embryo culture, surgical transfer to pseudopregnant females), micromanipulation equipment, and molecular genetics. The first application of mouse embryo micromanipulation was to test the developmental capacity of blastomeres [88]. In this experiment, blastomeres in two-cell and four-cell embryos were lysed by using micromanipulation, and the remaining blastomeres were tested for developmental potential after implantation into pseudopregnant recipients. The first mouse zygote microinjection experiment tested whether eggs injected with a gamma globulin solution would survive and develop into mice after transfer to recipients [16]. Other applications of microinjection included experiments on oocyte biology [89], the study of preimplantation developmental biology [90], and the translation of exogenous mRNA species [91]. By the time that molecular biology established DNA cloning as a standard method, all other necessary elements for the production of transgenic mice were in place. Following the initial report of transgenic mice, the field grew rapidly (Table 2).

6.3.2 Technique Development

The microinjection of DNA into the mouse pronucleus is typically carried out on a research microscope providing 400× magnification. The zygotes are placed in an injection chamber, which may

Table 2
Publication history relevant to genetically engineered mouse models

Publication year	Transgenic mouse papers	Mouse ES cell papers	Endonuclease-mediated mouse papers
1980	1	48	
1981	1	69	
1982	3	85	
1983	8	87	
1984	17	117	
1985	55	141	
1986	65	147	
1987	141	148	
1988	243	153	
1989	428	179	
1990	534	175	
1991	743	191	
1992	825	221	
1993	1171	262	
1994	1486	309	
1995	2054	289	
1996	2617	334	
1997	3332	404	
1998	4129	437	
1999	5113	428	
2000	6170	503	
2001	7010	549	
2002	7786	641	
2003	8195	737	
2004	9166	834	
2005	9549	943	
2006	10,314	1031	
2007	11,065	1058	
2008	11,769	1094	3
2009	12,449	1101	4
2010	13,211	1218	9

(continued)

Table 2
(continued)

Publication year	Transgenic mouse papers	Mouse ES cell papers	Endonuclease-mediated mouse papers
2011	13,998	1238	15
2012	14,389	1138	19
2013	14,715	1235	66
2014	14,369	1305	153
2015	13,939	1342	255
2016	13,825	1310	482
2017	12,748	1135	683
2018	6892	893	674

Results obtained from https://www.ncbi.nlm.nih.gov/pubmed. Number of publications found in PubMed when the PubMed database is searched with the term "Transgenic Mice" (results given in the column labeled Transgenic Mouse Papers). Results from the search term "Mouse ES Cells" are given in the column labeled Mouse ES Cell Papers. Results for the search terms "Zinc Finger Nucleases AND Mice," "TALENs AND Mice," and "CRISPR AND MICE" are combined in the column labeled Endonuclease-mediated mouse papers

consist of a hanging drop slide, depression slide, petri dish, or other arrangement that provides a medium compatible with mouse zygote cultures such as M2 [45] and is protected from evaporation by a layer of mineral oil. A glass holding pipette with an outer diameter of 120 μm and an inner diameter of 10–30 μm is connected to a piston syringe that is filled with air or a nonmiscible, nonaqueous medium. The piston syringe is used to apply suction to the pipette and thus draw the fertilized egg onto the holding pipette and hold it still. Alternatively, a mouth-controlled pipette can be used to apply suction and positive pressure to hold and release the zygote. The pronuclear membrane is then brought into focus by raising or lowering the holding pipette with the micromanipulator. Opposite to the zygote is a microinjection needle with an opening of less than 1 μm in diameter, a diameter that cannot be resolved by light microscopy. The needle is brought into the same plane of focus as the pronuclear membrane within the zygote with the micromanipulator. The tip of the needle is guided through the zona pellucida, the cell membrane, and the pronuclear membrane. Pressure is applied to the DNA solution in the needle by pneumatic pressure or by a piston syringe driving a nonmiscible, nonaqueous solution against the DNA within the microinjection needle. DNA delivery to the pronucleus is visually observed as an expansion in the size of the pronucleus that indicates DNA solution has been introduced. The volume administered by this method has been estimated to contain approximately three picoliters and approximately 300 plasmid DNA transgene molecules or a dozen bacterial artificial chromosome (BAC) transgene molecules [80]. After microinjection, the zygotes are placed in culture medium in an incubator until the time comes to transfer them to the oviduct of a pseudopregnant recipient female.

6.3.3 Mechanism of Integration

Small plasmid transgenes integrate more efficiently when they are introduced as linear instead of circular molecules [80, 92]. Initial studies indicated that transgene DNA molecules were present as head-to-tail concatemers in transgenic mice after microinjection [92, 93]. This led to a general model of exogenous DNA integration into chromosomes that supposed that circularly permuted transgene DNA molecules underwent homologous recombination in the cell and concatemer formation prior to DNA integration on a random chromosome [94, 95].

The advent of genomic DNA sequencing technologies gave evidence that although the basic outline of transgene integration structures is correct (multicopy arrays are common), the structure of the array is more complex than was suggested by the earlier studies [96–100]. These results are consistent with experiments on large transgenes such as yeast artificial chromosomes (YACs) and bacterial artificial chromosomes (BACs), which indicate that integrations can include internal transgene rearrangements in transgenic mice [101, 102]. Genomic sequencing data show that transgenes, whether small plasmid transgenes or larger transgenes such as BACs, integrate into the chromosome in a process that is reminiscent of chromothripsis or "chromosome shattering" [97]. These results indicate that it is common for small (plasmid) and large (BAC) transgenes to undergo fragmentation and rearrangement during transgene integration and are incongruent with simpler head-to-tail concatemer models that were proposed prior to the availability of genome sequencing tools. In practice, multiple transgenic founders should be analyzed for transgene expression and phenotypic consequences to control for random transgene integrations that can disrupt endogenous genes and produce confounding phenotypes [103, 104].

6.4 Transgenic Mice: ES Cell-Mediated Transgenesis

6.4.1 Targeted Mutation

In contrast to the random transgene integration events that occur when DNA is microinjected into a zygote, with the accompanying random chromosomal integration and potential for genetic mutation, the use of homologous recombination in mouse embryonic stem (ES) cells permits the precise targeting of genetic changes to specific places in the genome. The critical abilities required for this procedure are the culture of germline competent ES cells, the genetic modification of ES cells with gene-targeting vectors, screening and identification of genetically engineered ES cells, and the subsequent use of the targeted euploid ES cells to produce ES cell-mouse chimeras that will transmit the desired mutation through the germline.

6.4.2 Development of Technologies

In 2007, the Nobel Prize in Physiology or Medicine was awarded to Mario Capecchi, Martin Evans, and Oliver Smithies for their contributions to ES cell technology. Dr. Smithies established that homologous recombination between synthetic vectors and chromosomes in cells was possible [105], Dr. Evans identified ES cells [106] and Dr. Capecchi demonstrated that specific gene modifications could be introduced in mice by using homologous recombination in mouse ES cells [107]. The first germline ES cell-mouse chimeras were reported in 1984 [108], followed by targeted gene replacement in ES cells [109, 110], and then followed by germline transmission of gene-targeted ES cells in 1989 [111, 112]. There quickly followed a number of refinements to the basic process of gene targeting in ES cells. The method for positive/negative selection with neomycin phosphotransferase (confers resistance to G418) and herpes simplex virus thymidine kinase (confers sensitivity to ganciclovir and FIAU: 1-(2-deoxy-2-fluoro-beta-D-arabinofuranosyl)-5-iodouracil) was described by Mansour and coworkers [113]. The promoter from the mouse phosphoglycerate kinase I gene [114] was shown to be more effective than viral promoters for driving G418 resistance in ES cells [115]. The PGK-neo cassette is a standard tool used to generate gene-targeted ES cells for the Knockout Mouse Project intended to knock out all the genes in the mouse genome in ES cells (www.komp.org), [116]. The use of the diphtheria toxin A gene for negative selection is more effective than the HSV-TK and has been adopted by KOMP for its gene-targeting vectors [116, 117]. The dependence on isogenic DNA in gene-targeting vectors for efficient homologous recombination with genomic DNA targets in mouse ES cells was recognized in 1992 [118]. The development of robust ES cells that contributed to all tissues in a fetus after combination with tetraploid blastomeres also made gene-targeting experiments very efficient [119]. Following these refinements in ES cell technology, the number of research papers on ES cells and gene-targeted mice rapidly escalated (Table 2).

6.4.3 Chimeric Mouse Production

Unlike DNA microinjection, methods for the introduction or combination of mouse ES cells and blastocysts, morula, or other embryo stages have undergone considerable changes since the initial method was described [120]. This first approach to blastocyst microinjection involved the use of three recurved needles to hold open a slit made in the blastocyst (Fig. 1). A simpler method using a single injection was described in 1972 that used a holding pipette in combination with a microinjection needle that had a beveled tip for ease of penetration into the blastocyst [121]. This method is essentially the same one followed by many transgenic core facilities. A third method involves the use of a blunt injection needle that is applied to the blastocyst by a rapid motion to pierce the zona pellucida and to introduce the needle tip into the

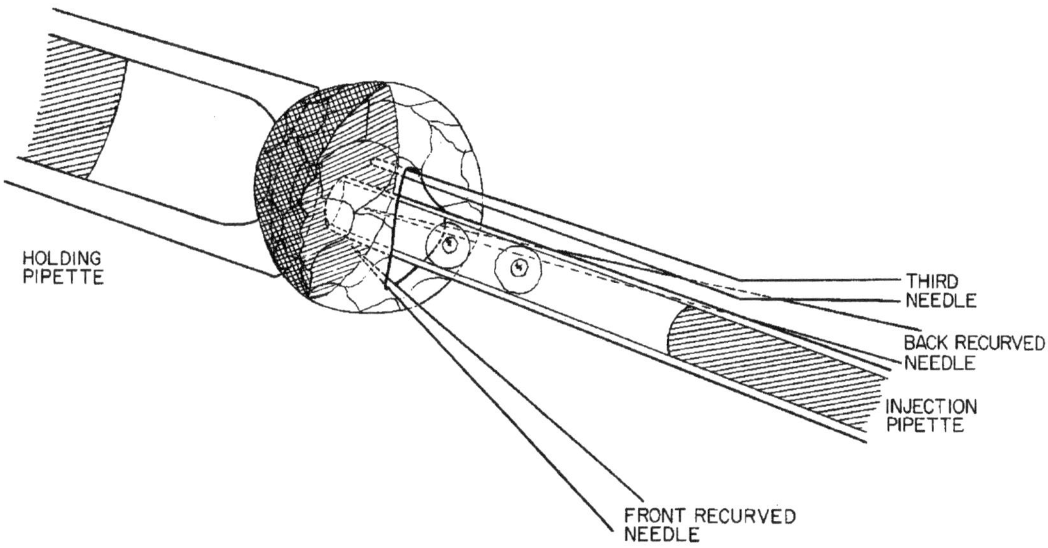

Fig. 1 Schematic representation of blastocyst microinjection technique developed by Richard Gardner. Note that the blastocyst is positioned on the holding pipet so that the inner cell mass is adjacent to the holding pipette. Two fine glass needles are placed in the blastocyst and used to produce a slit by scissor motion from a double instrument head mounted on a Leitz micromanipulator. A third needle is introduced to hold open the slit, while a third micromanipulator is used to position the injection pipette within the blastocyst and to eject cells into the blastocyst (Figure used with permission from Gardner (1968) [120])

blastocoel (Fig. 2), [122]. A variation on blastocyst microinjection to improve the efficiency of the process is to use a laser to create a slice in the zona pellucida and trophoblast cells to facilitate the penetration of a beveled injection needle for the delivery of ES cells to the blastocoel [123]. It has also been found that ES cells can be combined with mouse embryo developmental stages besides the blastocyst. One common stage that is used for microinjection is the uncompacted morula [123, 124]; however, earlier stages of the preimplantation embryo can also be successfully microinjected with ES cells for chimera production [125–127]. Besides the microinjection approach, many laboratories use aggregation methods to combine ES cells with uncompacted morula for the production of chimeras [128, 129].

6.5 Endonuclease-Mediated Genetic Engineering

In the current era, CRISPR/Cas9 technology has dramatically changed the landscape of transgenic animal model engineering. CRISPR/Cas9 technology and its precursors (meganucleases, zinc finger nucleases (ZFN), and TALE nucleases) all rely on the same central principle: (1) a chromosome break is produced that is followed by (2) repair of the break by either a nonhomologous end-joining event (NHEJ) or a homology-directed repair event (HDR). In the former case, chromosome breaks are repaired and result in small insertions or deletions that will alter the original sequence of the chromosome. If this occurs within a coding exon of

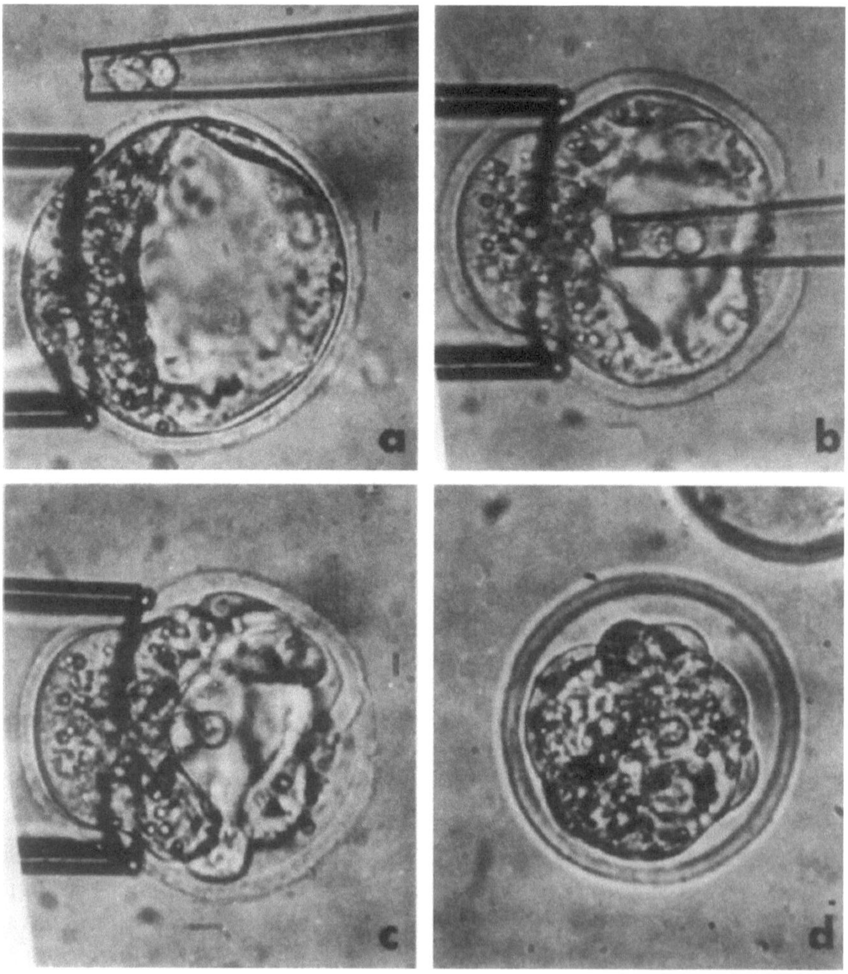

Fig. 2 Blastocyst microinjection technique developed by Charles Babinet. The injection needle is positioned so that it is in the same focal plane as the blastocyst. (**a**) The micromanipulator with the injection needle is adjusted so that the needle cannot move past the inner cell mass on the left side. (**b**) After ES cells are loaded and the needle is positioned, a quick thrust is used to rapidly move the blunt injection needle into the blastocoel. Cells are then microinjected into the cavity of the blastocyst. (**c**) Cells within the blastocyst. (**d**) Collapsed blastocyst after microinjection (Figure used with permission from Babinet (1980) [122])

a gene, it can lead to a gene knockout or to a change in a protein's amino acid sequence. Homology-directed repair occurs when a DNA donor molecule, which may be an oligonucleotide, a double-stranded DNA plasmid donor, or a long single-stranded DNA donor, is integrated into the chromosome in a precise fashion. The use of ES cell technology for precise mutations in the mouse genome has largely been supplanted by the use of CRISPR/Cas9-mediated HDR. CRISPR/Cas9 is especially appealing for other species such as the rat, in which the use of ES cell technology is very challenging, and for species in which ES cells with germline potential are unavailable.

6.5.1 Other Nuclease Technologies

Genome modifications that use meganucleases require significant protein engineering to modify the meganuclease structure to specifically target the DNA sequence of interest. This has limited the applications of meganucleases to the production of transgenic mice and rats although it is still possible to produce transgenic rats and mice via this technology [130]. It is somewhat less challenging to obtain specific zinc finger nucleases that display high chromosome cutting activity. The first demonstrations of ZFN knockout and knockin rat and mouse models were in 2009–2010 [131–134]. The relative ease with which ZFN can be used to target the rat genome led the National Heart, Lung, and Blood Institute (NIH) to establish the PhysGen Knockout program to generate new genetically engineered rat models [135]. Commercially available ZFN preparations are costly (Sangamo Therapeutics), although there is a ZFN assembly kit available at Addgene.org [136]. By comparison, the effort to generate ZFNs in house greatly exceeds the effort needed to prepare CRISPR/Cas9 reagents. As a result of these factors, ZFN technology is not routinely applied to the generation of new genetically engineered mouse models. TALENs (transcription activator-like effector nucleases) were first applied to produce genetically engineered rat models in 2011 [137], which led to an explosion of mouse models produced with TALENs in 2013 [138–143]. Commercially available TALEN reagents also proved to be costly (Cellectis SA), although TALEN assembly kits such as the one described by Cermak and colleagues [144] are available at Addgene.org. While TALENs are simpler to construct than ZFN or meganucleases, they are less active in producing chromosome breaks in mouse and rat zygotes than CRISPR reagents (unpublished observations).

6.5.2 CRISPR/Cas9

The mechanism of action of both ZFN and TALENs relies on the endonuclease activity of FokI endonuclease domains [145]. Both ZFN and TALENs have now been supplanted by CRISPR/Cas9 technology due to superior ease of use and flexibility in targeting the genome with a synthetic RNA molecule. After the demonstration that CRISPR/Cas9 could be used to target chromosomes in mammalian cell lines in 2013 [146–148], the field rapidly applied this technology to the genetic engineering of mice and rats (Table 2). CRISPR/Cas9 reagents can be microinjected into mouse eggs in the form of (1) plasmids that co-express Cas9 and guide RNA molecules, (2) mRNA coding for Cas9 plus guide RNA molecules, or (3) Cas9 protein complexed with guide RNA. All of these delivery forms are effective; however, the current consensus in the field is that Cas9 ribonuclease protein complexes offer advantages because they avoid random integration of plasmids in the genome (which can be problematic in later generations of animals) and the immediate activity of Cas9 on the genome reduces mosaicism in founder animals. Indeed, CRISPR/Cas9 targeting is so efficient that one can virtually guarantee gene knockouts by the introduction of premature termination codons in coding genes after indel formation [149–151] or the excision of chromosome regions [152–154]. Similarly, the introduction of point mutations or epitope tags by HDR with oligonucleotides is so efficient that one can almost guarantee success [155–157]. An early publication that excited great attention was the use of CRISPR/Cas9 to knock in reporter molecules and generate mice with conditional (floxed) genes [158]. Subsequent research on improving the efficiency of reporter knockins and producing floxed alleles has evolved away from the methods in this initial report. For example, reporter knockins can be achieved with circular DNA plasmid donors [158–160]; however, the efficiency is much higher when CRISPR/Cas9 cuts on the chromosome are combined with long single-stranded DNA molecules [161–163] or when linear double-stranded DNA templates are microinjected into one- or two-cell fertilized mouse eggs [164–166]. In particular, the efficiency of producing floxed alleles with a long single-stranded DNA template has been shown to be much more efficient than procedures involving microinjection of two CRISPR reagents and two oligonucleotides designed to introduce loxP sites around a critical exon [167–170]. One variable that is under consideration in the field is the length of the arms of homology in the DNA donor. It was shown that a simple PCR product with 36 base pair arms of homology flanking mCherry is sufficient to result in knockins by HDR [165], although the tendency in the field is to use arms of homology between 1 and 2 kb in length [164, 171, 172]. In general, the principle of introducing exogenous coding sequences such as fluorescent protein reporters or recombinases in gene-targeting vectors still follows principles of designing gene-targeted ES cells [173–175]. On either side of the sequence to be inserted, a 5′ and 3′ arm of homology is required to direct the incorporation of the synthetic DNA into the desired locus. The differences are that one does not need to include a drug selection marker nor a negative selection marker such as the diphtheria toxin A chain in CRISPR/Cas9 gene-targeting vectors.

6.5.3 Minimizing Off Target Mutations

The use of CRISPR/Cas9 technology introduces the need to control for off target mutations in the genome. It should be noted that CRISPR-Cas9 off target hits have been observed in mouse and rat zygotes and in mouse ES cells [176, 177], a problem that is infrequently observed with conventional ES cell technology. This is noteworthy, as it is important to control for off target hits so that phenotypes observed in model organisms can be attributed to the engineered mutation. The first step is in the selection of the guide RNA sequence. Ideally, a highly specific guide is selected which lacks possible off target hits in coding exons and, when possible, within introns. The computational website at crispor.tefor.net offers tools that facilitate the identification of such guide RNA targets [178]. After the selection of the guide RNA, the next element under the control of the experimentalist is the choice of Cas9 enzyme. Since the initial work with the wild-type Cas9 enzyme was published, a multitude of variants have been described. Of particular interest are those that reduce the probability of off target hits. Enhanced specificity Cas9 [179] and high-fidelity Cas9 [180] were demonstrated to significantly reduce off target hits when used to produce genome-edited mice and rats [176]. For those laboratories who do not include recombinant protein preparation in their repertoires, there are at least two commercial sources of Cas9 protein to consider: the eSpCas9 protein offered by MilliporeSigma [179] and the HiFi Cas9 protein offered by IDTDNA [181]. When analysis of off target effects in animals beyond the founder generation is to be undertaken, it is essential to conduct the studies carefully so as to avoid confounding effects from normal genetic drift due to meiotic crossing over that naturally occurs during development [176, 182, 183]. In general, when a highly specific guide is combined with a highly specific Cas9 protein, off target hits are minimized. Analysis of mice from more than one founder should display the same phenotype unless a spontaneous mutation has occurred that affects the mice [184, 185].

7 Future Directions

The application of CRISPR/Cas9 to the generation of genetically engineered mice and rats has had an immense impact on the ease with which animal models can be generated for research. CRISPR/Cas9 technology may be supplanted at some time in the future by another technology such as a DNA-guided endonuclease (*see* **Note 2**). Initially, it was hoped that the *Natronobacterium gregoryi* Argonaute enzyme might be such a tool, but it did not function as desired [186]. Another alternative to Cas9 is the Cpf1 enzyme that has been successfully applied to the generation of a mouse model [187, 188]. However, it seems that Cpf1 is not as efficient as Cas9 and the protospacer adjacent motif TTTN of four nucleotides required by Cpf1 limits the number of possible genomic targets. It may be that sometime in the future, an even more active and specific endonuclease will become available for genome editing. Until that time, researchers continue to marvel at the effectiveness of Cas9 and to optimize its efficiency.

8 Notes

1. Besides genetic background, there are many other factors that can affect the efficiency of transgenic founder production, including use of linear or circular transgene DNA, purification of DNA, microinjection needle geometry, volume of microinjected DNA solution, media and incubation conditions for zygotes, sensitivity of transgene detection methods, mouse colony health status, nutritional status of pseudopregnant females, and technical skills of transgenic core facility personnel, among other factors [43, 62, 75, 80, 189, 190].
2. Due to a lack of space, a number of technologies (transposases, integrases, etc.) and refinements of Cas9 delivery (electroporation of embryos and gonads) were not discussed. The reader is invited to seek elsewhere for the latest information regarding these developments.

References

1. Gates WH (1925) The Japanese waltzing mouse, its origin and genetics. Proc Natl Acad Sci U S A 11:651–653
2. Crow JF (2002) C. C. Little, cancer and inbred mice. Genetics 161:1357–1361
3. Morse HCIII (ed) (1978) Origins of inbred mice. Academic Press, New York
4. Silver LM (1995) Mouse genetics: concepts and applications. Oxford University Press, Oxford
5. Silvers WK (1979) The coat colors of mice: a model for mammalian gene action and interaction. Springer Verlag, Berlin
6. Russell LB (2013) The mouse house: a brief history of the ORNL mouse-genetics program, 1947–2009. Mutat Res 753 (2):69–90. PMID: 23994540
7. Watson JD, Crick FH (1953) Molecular structure of nucleic acids; a structure for deoxyribose nucleic acid. Nature 171:737–738
8. Jackson DA, Symons RH, Berg P (1972) Biochemical method for inserting new genetic information into DNA of simian virus 40: circular SV40 DNA molecules containing lambda phage genes and the galactose operon of Escherichia coli. Proc Natl Acad Sci U S A 69:2904–2909
9. Cobb RE, Ning JC, Zhao H (2014) DNA assembly techniques for next-generation combinatorial biosynthesis of natural products. J Ind Microbiol Biotechnol 41 (2):469–477. PMID: 24127070
10. Hughes RA, Miklos AE, Ellington AD (2011) Gene synthesis: methods and applications. Methods Enzymol 498:277–309
11. Summers MC (2014) A brief history of the development of the KSOM family of media. Hum Fertil (Camb) 17(Suppl 1):12–16
12. Zarrow MX, Wilson ED (1961) The influence of age on superovulation in the immature rat and mouse. Endocrinology 69:851–855
13. Spearow JL (1988a) Major genes control hormone-induced ovulation rate in mice. J Reprod Fertil 82:787–797
14. Nagy A, Gertsenstein M, Vintersten K, Behringer R (2014) Manipulating the mouse embryo: a laboratory manual, 4th edn. Cold Spring Harbor Laboratory Press, Cold Spring Harbor, New York
15. El-Badry HM (1963) Micromanipulators and micromanipulation. Springer Verlag, Vienna
16. Lin TP (1966) Microinjection of mouse eggs. Science 151:333–337
17. Gordon JW, Scangos GA, Plotkin DJ, Barbosa JA, Ruddle FH (1980) Genetic transformation of mouse embryos by microinjection of purified DNA. Proc Natl Acad Sci U S A 77:7380–7384
18. Aschheim Z (1928) Die schwangerschaftsdiagnose aus dem harn durch nachweis des hypophysenvorderlappenhormons. Klin Wochenschr 7:1404–1411
19. Cole HH, Hart GH (1930) The potency of blood serum of mares in progressive stages of

pregnancy in effecting the sexual maturity of the immature rat. Am J Phys 93:57–68

20. Evans HM, Gustus EL, Simpson ME (1933) Concentration of the gonadotropic hormone in pregnant mare's serum. J Exp Med 58:569–574
21. Runner MN, Gates A (1954) Conception in prepuberal mice following artificially induced ovulation and mating. Nature 174:222–223
22. Runner MN, Palm J (1953) Transplantation and survival of unfertilized ova of the mouse in relation to postovulatory age. J Exp Zool 134:303–316
23. Van Blerkom J, Runner MN (1976) The fine structural development of preimplantation mouse parthenotes. J Exp Zool 196:113–124
24. Wilson ED, Zarrow MX (1962) Comparison of superovulation in the immature mouse and rat. J Reprod Fertil 3:148–158
25. Gates AH, Bozarth JL (1978) Ovulation in the PMSG-treated immature mouse: effect of dose, age, weight, puberty, season and strain (BALB/c, 129 and C129F1 hybrid). Biol Reprod 18:497–505
26. Cosby NC, Chou K, Dukelow WR (1989) Embryo production in B6D2-F1 mice using two superovulating regimens. Lab Anim Sci 39:249–250
27. Taketo M, Schroeder AC, Mobraaten LE, Gunning KB, Hanten G, Fox RR, Roderick TH, Stewart CL, Lilly F, Hansen CT, Overbeek PA (1991) FVB/N: an inbred mouse strain preferable for transgenic analyses. Proc Natl Acad Sci U S A 88(6):2065–2069. PMID: 1848692
28. Brooke DA, Orsi NM, Ainscough JF, Holwell SE, Markham AF, Coletta PL (2007) Human menopausal and pregnant mare serum gonadotrophins in murine superovulation regimens for transgenic applications. Theriogenology 67:1409–1413
29. Popova E, Krivokharchenko A, Ganten D, Bader M (2002) Comparison between PMSG- and FSH induced superovulation for the generation of transgenic rats. Mol Reprod Dev 63:177–182
30. Filipiak WE, Saunders TL (2006) Advances in transgenic rat production. Transgenic Res 15:673–686
31. Takeo T, Nakagata N (2015) Superovulation using the combined administration of inhibin antiserum and equine chorionic gonadotropin increases the number of ovulated oocytes in C57BL/6 female mice. PLoS One 10: e0128330
32. Hasegawa A, Mochida K, Inoue H, Noda Y, Endo T, Watanabe G, Ogura A (2016) High-yield superovulation in adult mice by anti-inhibin serum treatment combined with estrous cycle synchronization. Biol Reprod 94:21
33. Nakagawa Y, Sakuma T, Nishimichi N, Yokosaki Y, Yanaka N, Takeo T, Nakagata N, Yamamoto T (2016) Ultra-superovulation for the CRISPR-Cas9-mediated production of gene-knockout, single-amino-acid-substituted, and floxed mice. Biol Open 5 (8):1142–11428
34. Runner MN (1947) Development of mouse eggs in the anterior chamber of the eye. Anat Rec 98:1–17
35. Hammond J Jr (1949) Recovery and culture of tubal mouse ova. Nature 163:28–29
36. Whitten WK (1956) Culture of tubal mouse ova. Nature 177:96
37. Whitten WK (1957) Culture of tubal ova. Nature 179:1081–1082
38. Brinster RL (1965) Studies on the development of mouse embryos in vitro. IV. Interaction of energy sources. J Reprod Fertil 10:227–240
39. Brinster RL, Biggers JD (1965) In-vitro fertilization of mouse ova within the explanted fallopian tube. J Reprod Fertil 10:277–279
40. Whitten WK, Biggers JD (1968) Complete development in vitro of the pre-implantation stages of the mouse in a simple chemically defined medium. J Reprod Fertil 17:399–401
41. Chatot CL, Ziomek CA, Bavister BD, Lewis JL, Torres I (1989) An improved culture medium supports development of random-bred 1-cell mouse embryos in vitro. J Reprod Fertil 86:679–688
42. Chatot CL, Lewis JL, Torres I, Ziomek CA (1990) Development of 1-cell embryos from different strains of mice in CZB medium. Biol Reprod 42:432–440
43. Erbach GT, Lawitts JA, Papaioannou VE, Biggers JD (1994) Differential growth of the mouse preimplantation embryo in chemically defined media. Biol Reprod 50 (5):1027–1033. PMID: 8025158
44. Biggers JD, McGinnis LK, Raffin M (2000) Amino acids and preimplantation development of the mouse in protein-free potassium simplex optimized medium. Biol Reprod 63:281–293
45. Quinn P, Barros C, Whittingham DG (1982) Preservation of hamster oocytes to assay the fertilizing capacity of human spermatozoa. J Reprod Fertil 66:161–168
46. Whittingham DG (1971) Culture of mouse ova. J Reprod Fertil Suppl 14:7–21

47. Lawitts JA, Biggers JD (1993) Culture of preimplantation embryos. Methods Enzymol 225:153–164
48. Biggers JD (1998) Reflections on the culture of the preimplantation embryo. Int J Dev Biol 42:879–884
49. Summers MC, Biggers JD (2003) Chemically defined media and the culture of mammalian preimplantation embryos: historical perspective and current issues. Hum Reprod Update 9:557–582
50. Heape W (1890) Preliminary note on the transplantation and growth of mammalian ova within a uterine foster mother. Proc Roy Soc London B 48:457–458
51. Nicholas JS, Rudnick D (1934) The development of rat embryos in tissue culture. Proc Natl Acad Sci U S A 20:656–658
52. Bittner JJ, Little CC (1937) Transmission of breast and lung cancer in mice. J Hered 28:117–121
53. McLaren A, Michie D (1956) Studies on the transfer of fertilized mouse eggs to uterine foster-mothers. J Exp Biol 33:394–416
54. Fekete E, Little CC (1942) Observations on the mammary tumor incidence of mice born from transferred ova. Cancer Res 2:525–530
55. Tarkowski AK (1959b) Experiments on the transplantation of ova in mice. Acta Theriol 2:251–267
56. Ueda O, Yorozu K, Kamada N, Jishage K, Kawase Y, Toyoda Y, Suzuki H (2003) Possible expansion of "window of implantation" in pseudopregnant mice: time of implantation of embryos at different stages of development transferred into the same recipient. Biol Reprod 69:1085–1090
57. Goto Y, Noda Y, Shiotani M, Kishi J, Nonogaki T, Mori T (1993) The fate of embryos transferred into the uterus. J Assist Reprod Genet 10:197–201
58. Johnson LW, Moffatt RJ, Bartol FF, Pinkert CA (1996) Optimization of embryo transfer protocols for mice. Theriogenology 46:1267–1276
59. Chin HJ, Wang CK (2001) Utero-tubal transfer of mouse embryos. Genesis 30:77–81
60. Pease S, Schroeder AC, Schmidt GH (1989) Production of transgenic mice: acupuncture needle-facilitated embryo transfer to oviduct ampulla. Trends Genet 5:293
61. Zhang Z, Lv X, Wang Y, Chen Y, Zheng R, Sun H, Bian G, Xiao Y, Li Q, Yang Q, Ai J, Duan J, Tan R, Liu Y, Yang Y, Wei Y, Zhou Q (2009) Success of murine embryo transfer increased by a modified transfer pipette. J Reprod Dev 55:94–97
62. Fielder TJ, Montoliu L (2011) Transgenic production benchmarks. In: Pease S, Saunders TL (eds) Advanced protocols for animal transgenesis: an ISTT manual. Springer-Verlag, Berlin, pp 81–97
63. Gates AH (1956) Viability and developmental capacity of eggs from immature mice treated with gonadotrophins. Nature 177:754–755
64. Bronson RA, McLaren A (1970) Transfer to the mouse oviduct of eggs with and without the zona pellucida. J Reprod Fertil 22:129–137
65. Rülicke T, Haenggli A, Rappold K, Moehrlen U, Stallmach T (2006) No transuterine migration of fertilised ova after unilateral embryo transfer in mice. Reprod Fertil Dev 18:885–891
66. Beatty RA (1951) Transplantation of mouse eggs. Nature 168:995
67. Cui L, Zhang Z, Sun F, Duan X, Wang M, Di K, Li X (2014) Transcervical embryo transfer in mice. J Am Assoc Lab Anim Sci 53:228–231
68. Green M, Bass S, Spear B (2009) A device for the simple and rapid transcervical transfer of mouse embryos eliminates the need for surgery and potential post-operative complications. BioTechniques 47:919–924
69. Steele KH, Hester JM, Stone BJ, Carrico KM, Spear BT, Fath-Goodin A (2013) Nonsurgical embryo transfer device compared with surgery for embryo transfer in mice. J Am Assoc Lab Anim Sci 52:17–21
70. Spearow JL (1988b) Characterization of genetic differences in hormone-induced ovulation rate in mice. J Reprod Fertil 82:799–806
71. Legge M, Sellens MH (1994) Optimization of superovulation in the reproductively mature mouse. J Assist Reprod Genet 11:312–318
72. Roudebush WE, Duralia DR (1996) Superovulation, fertilization, and in vitro embryo development in BALB/cByJ, BALB/cJ, B6D2F1/J, and CFW mouse strains. Lab Anim Sci 46:239–240
73. Osman GE, Jacobson DP, Li SW, Hood LE, Liggitt HD, Ladiges WC (1997) SWR: an inbred strain suitable for generating transgenic mice. Lab Anim Sci 47:167–171
74. Vergara GJ, Irwin MH, Moffatt RJ, Pinkert CA (1997) In vitro fertilization in mice: strain differences in response to superovulation protocols and effect of cumulus cell removal. Theriogenology 47:1245–1252
75. Auerbach AB, Norinsky R, Ho W, Losos K, Guo Q, Chatterjee S, Joyner AL (2003)

Strain-dependent differences in the efficiency of transgenic mouse production. Transgenic Res 12:59–69
76. Byers SL, Payson SJ, Taft RA (2006) Performance of ten inbred mouse strains following assisted reproductive technologies (ARTs). Theriogenology 65:1716–1726
77. Luo C, Zuñiga J, Edison E, Palla S, Dong W, Parker-Thornburg J (2011) Superovulation strategies for 6 commonly used mouse strains. J Am Assoc Lab Anim Sci 50:471–478
78. Mochida K, Hasegawa A, Otaka N, Hama D, Furuya T, Yamaguchi M, Ichikawa E, Ijuin M, Taguma K, Hashimoto M, Takashima R, Kadota M, Hiraiwa N, Mekada K, Yoshiki A, Ogura A (2014) Devising assisted reproductive technologies for wild-derived strains of mice: 37 strains from five subspecies of Mus musculus. PLoS One 9:e114305
79. Kolbe T, Landsberger A, Manz S, Na E, Urban I, Michel G (2015) Productivity of superovulated C57BL/6J oocyte donors at different ages. Lab Anim (NY) 44:346–349
80. Brinster RL, Chen HY, Trumbauer ME, Yagle MK, Palmiter RD (1985) Factors affecting the efficiency of introducing foreign DNA into mice by microinjecting eggs. Proc Natl Acad Sci U S A 82:4438–4442
81. Fielder TJ, Barrios L, Montoliu L (2010) A survey to establish performance standards for the production of transgenic mice. Transgenic Res 19:675–681
82. Berg P, Mertz JE (2010) Personal reflections on the origins and emergence of recombinant DNA technology. Genetics 184:9–17
83. Cohen SN (2013) DNA cloning: a personal view after 40 years. Proc Natl Acad Sci U S A 110:15521–15529
84. Bacchetti S, Graham FL (1977) Transfer of the gene for thymidine kinase to thymidine kinase-deficient human cells by purified herpes simplex viral DNA. Proc Natl Acad Sci U S A 74:1590–1594
85. Maitland NJ, McDougall JK (1977) Biochemical transformation of mouse cells by fragments of herpes simplex virus DNA. Cell 11:233–241
86. Wigler M, Silverstein S, Lee LS, Pellicer A, Cheng Y, Axel R (1977) Transfer of purified herpes virus thymidine kinase gene to cultured mouse cells. Cell 11:223–232
87. Graham FL, van der Eb AJ (1973) A new technique for the assay of infectivity of human adenovirus 5 DNA. Virology 52:456–467
88. Tarkowski AK (1959) Experiments on the development of isolated blastomeres of mouse eggs. Nature 184:1286–1287
89. Fulton BP, Whittingham DG (1978) Activation of mammalian oocytes by intracellular injection of calcium. Nature 273:149–151
90. Wilson IB, Bolton E, Cuttler RH (1972) Preimplantation differentiation in the mouse egg as revealed by microinjection of vital markers. J Embryol Exp Morphol 27:467–469
91. Brinster RL, Chen HY, Trumbauer ME, Avarbock MR (1980) Translation of globin messenger RNA by the mouse ovum. Nature 283:499–501
92. Folger KR, Wong EA, Wahl G, Capecchi MR (1982) Patterns of integration of DNA microinjected into cultured mammalian cells: evidence for homologous recombination between injected plasmid DNA molecules. Mol Cell Biol 2:1372–1387
93. McFarlane M, Wilson JB (1996) A model for the mechanism of precise integration of a microinjected transgene. Transgenic Res 5:171–177
94. Smith K (2001) Theoretical mechanisms in targeted and random integration of transgene DNA. Reprod Nutr Dev 41:465–485
95. Nakanishi T, Kuroiwa A, Yamada S, Isotani A, Yamashita A, Tairaka A, Hayashi T, Takagi T, Ikawa M, Matsuda Y, Okabe M (2002) FISH analysis of 142 EGFP transgene integration sites into the mouse genome. Genomics 80:564–557
96. Le Saux A, Houdebine LM, Jolivet G (2010) Chromosome integration of BAC (bacterial artificial chromosome): evidence of multiple rearrangements. Transgenic Res 19:923–931
97. Chiang C, Jacobsen JC, Ernst C, Hanscom C, Heilbut A, Blumenthal I, Mills RE, Kirby A, Lindgren AM, Rudiger SR, McLaughlan CJ, Bawden CS, Reid SJ, Faull RL, Snell RG, Hall IM, Shen Y, Ohsumi TK, Borowsky ML, Daly MJ, Lee C, Morton CC, MacDonald ME, Gusella JF, Talkowski ME (2012) Complex reorganization and predominant non-homologous repair following chromosomal breakage in karyotypically balanced germline rearrangements and transgenic integration. Nat Genet 44:390–397
98. Zhang R, Yin Y, Zhang Y, Li K, Zhu H, Gong Q, Wang J, Hu X, Li N (2012) Molecular characterization of transgene integration by next-generation sequencing in transgenic cattle. PLoS One 7:e50348
99. Dubose AJ, Lichtenstein ST, Narisu N, Bonnycastle LL, Swift AJ, Chines PS, Collins FS (2013) Use of microarray hybrid capture and next-generation sequencing to identify the anatomy of a transgene. Nucleic Acids Res 41:e70

100. Srivastava A, Philip VM, Greenstein I, Rowe LB, Barter M, Lutz C, Reinholdt LG (2014) Discovery of transgene insertion sites by high throughput sequencing of mate pair libraries. BMC Genomics 15:367
101. Smith DJ, Zhu Y, Zhang J, Cheng JF, Rubin EM (1995) Construction of a panel of transgenic mice containing a contiguous 2-Mb set of YAC/P1 clones from human chromosome 21q22.2. Genomics 27:425–434
102. Deal KK, Cantrell VA, Chandler RL, Saunders TL, Mortlock DP, Southard-Smith EM (2006) Distant regulatory elements in a Sox10-betaGEO BAC transgene are required for expression of Sox10 in the enteric nervous system and other neural crest-derived tissues. Dev Dyn 235:1413–1432
103. Meisler MH (1992) Insertional mutation of 'classical' and novel genes in transgenic mice. Trends Genet 8:341–344
104. Rijkers T, Peetz A, Rüther U (1994) Insertional mutagenesis in transgenic mice. Transgenic Res 3:203–215
105. Smithies O (2001) Forty years with homologous recombination. Nat Med 7:1083–1086
106. Evans M (2011) Discovering pluripotency: 30 years of mouse embryonic stem cells. Nat Rev Mol Cell Biol 12:680–686
107. Capecchi MR (2005) Gene targeting in mice: functional analysis of the mammalian genome for the twenty-first century. Nat Rev Genet 6:507–512
108. Bradley A, Evans M, Kaufman MH, Robertson E (1984) Formation of germ-line chimaeras from embryo-derived teratocarcinoma cell lines. Nature 309:255–256
109. Doetschman T, Gregg RG, Maeda N, Hooper ML, Melton DW, Thompson S, Smithies O (1987) Targeted correction of a mutant HPRT gene in mouse embryonic stem cells. Nature 330:576–578
110. Thomas KR, Capecchi MR (1987) Site-directed mutagenesis by gene targeting in mouse embryo-derived stem cells. Cell 51:503–512
111. Koller BH, Hagemann LJ, Doetschman T, Hagaman JR, Huang S, Williams PJ, First NL, Maeda N, Smithies O (1989) Germ-line transmission of a planned alteration made in a hypoxanthine phosphoribosyltransferase gene by homologous recombination in embryonic stem cells. Proc Natl Acad Sci U S A 86:8927–8931
112. Schwartzberg PL, Goff SP, Robertson EJ (1989) Germ-line transmission of a c-abl mutation produced by targeted gene disruption in ES cells. Science 246:799–803
113. Mansour SL, Thomas KR, Capecchi MR (1988) Disruption of the proto-oncogene int-2 in mouse embryo-derived stem cells: a general strategy for targeting mutations to non-selectable genes. Nature 336:348–352
114. Adra CN, Boer PH, McBurney MW (1987) Cloning and expression of the mouse pgk-1 gene and the nucleotide sequence of its promoter. Gene 60:65–74
115. Tybulewicz VL, Crawford CE, Jackson PK, Bronson RT, Mulligan RC (1991) Neonatal lethality and lymphopenia in mice with a homozygous disruption of the c-abl proto-oncogene. Cell 65:1153–1163
116. Skarnes WC, Rosen B, West AP, Koutsourakis M, Bushell W, Iyer V, Mujica AO, Thomas M, Harrow J, Cox T, Jackson D, Severin J, Biggs P, Fu J, Nefedov M, de Jong PJ, Stewart AF, Bradley A (2011) A conditional knockout resource for the genome-wide study of mouse gene function. Nature 474:337–342
117. Yagi T, Ikawa Y, Yoshida K, Shigetani Y, Takeda N, Mabuchi I, Yamamoto T, Aizawa S (1990) Homologous recombination at c-fyn locus of mouse embryonic stem cells with use of diphtheria toxin A-fragment gene in negative selection. Proc Natl Acad Sci U S A 87:9918–9922
118. te Riele H, Maandag ER, Berns A (1992) Highly efficient gene targeting in embryonic stem cells through homologous recombination with isogenic DNA constructs. Proc Natl Acad Sci U S A 89:5128–5132
119. Nagy A, Rossant J, Nagy R, Abramow-Newerly W, Roder JC (1993) Derivation of completely cell culture-derived mice from early-passage embryonic stem cells. Proc Natl Acad Sci U S A 90:8424–8428
120. Gardner RL (1968) The relationship between cell lineage and differentiation in the early mouse embryo. Results Probl Cell Differ 9:205–241
121. Moustafa LA, Brinster RL (1972) The fate of transplanted cells in mouse blastocysts in vitro. J Exp Zool 181:181–191
122. Babinet C (1980) A simplified method for mouse blastocyst injection. Exp Cell Res 130:15–19
123. Poueymirou WT, Auerbach W, Frendewey D, Hickey JF, Escaravage JM, Esau L, Doré AT, Stevens S, Adams NC, Dominguez MG, Gale NW, Yancopoulos GD, DeChiara TM, Valenzuela DM (2007) F0 generation mice fully derived from gene-targeted embryonic stem

cells allowing immediate phenotypic analyses. Nat Biotechnol 25:91–99

124. Tokunaga T, Tsunoda Y (1992) Efficacious production of viable germ-line chimeras between embryonic stem (ES) cells and eight-cell stage embryos. Develop Growth Differ 34:561–566
125. Huang J, Deng K, Wu H, Liu Z, Chen Z, Cao S, Zhou L, Ye X, Keefe DL, Liu L (2008) Efficient production of mice from embryonic stem cells injected into four- or eight-cell embryos by piezo micromanipulation. Stem Cells 26:1883–1890
126. De Repentigny Y, Kothary R (2010) Production of mouse chimeras by injection of embryonic stem cells into the perivitelline space of one-cell stage embryos. Transgenic Res 19:1137–1344
127. Kraus P, Leong G, Tan V, Xing X, Goh JW, Yap SP, Lufkin T (2010) A more cost effective and rapid high percentage germ-line transmitting chimeric mouse generation procedure via microinjection of 2-cell, 4-cell, and 8-cell embryos with ES and iPS cells. Genesis 48:394–349
128. Tanaka M, Hadjantonakis AK, Vintersten K, Nagy A (2009) Aggregation chimeras: combining ES cells, diploid, and tetraploid embryos. Methods Mol Biol 530:287–309
129. Williams E, Auerbach W, DeChiara TM, Gertsenstein M (2011) Combining ES cells with embryos. In: Pease S, Saunders TL (eds) Advanced protocols for animal transgenesis: an ISTT manual. Springer-Verlag, Berlin, pp 377–430
130. Ménoret S, Fontanière S, Jantz D, Tesson L, Thinard R, Rémy S, Usal C, Ouisse LH, Fraichard A, Anegon I (2013) Generation of Rag1-knockout immunodeficient rats and mice using engineered meganucleases. FASEB J 27:703–711
131. Geurts AM, Cost GJ, Freyvert Y, Zeitler B, Miller JC, Choi VM, Jenkins SS, Wood A, Cui X, Meng X, Vincent A, Lam S, Michalkiewicz M, Schilling R, Foeckler J, Kalloway S, Weiler H, Ménoret S, Anegon I, Davis GD, Zhang L, Rebar EJ, Gregory PD, Urnov FD, Jacob HJ, Buelow R (2009) Knockout rats via embryo microinjection of zinc-finger nucleases. Science 325:433
132. Mashimo T, Takizawa A, Voigt B, Yoshimi K, Hiai H, Kuramoto T, Serikawa T (2010) Generation of knockout rats with X-linked severe combined immunodeficiency (X-SCID) using zinc-finger nucleases. PLoS One 5:e8870
133. Meyer M, de Angelis MH, Wurst W, Kühn R (2010) Gene targeting by homologous recombination in mouse zygotes mediated by zinc-finger nucleases. Proc Natl Acad Sci U S A 107:15022–15026
134. Cui X, Ji D, Fisher DA, Wu Y, Briner DM, Weinstein EJ (2011) Targeted integration in rat and mouse embryos with zinc-finger nucleases. Nat Biotechnol 29:64–67
135. Geurts AM, Moreno C (2010) Zinc-finger nucleases: new strategies to target the rat genome. Clin Sci (Lond) 19:303–311
136. Wright DA, Thibodeau-Beganny S, Sander JD, Winfrey RJ, Hirsh AS, Eichtinger M, Fu F, Porteus MH, Dobbs D, Voytas DF, Joung JK (2006) Standardized reagents and protocols for engineering zinc finger nucleases by modular assembly. Nat Protoc 1:1637–1652
137. Tesson L, Usal C, Ménoret S, Leung E, Niles BJ, Remy S, Santiago Y, Vincent AI, Meng X, Zhang L, Gregory PD, Anegon I, Cost GJ (2011) Knockout rats generated by embryo microinjection of TALENs. Nat Biotechnol 29:695–696
138. Davies B, Davies G, Preece C, Puliyadi R, Szumska D, Bhattacharya S (2013) Site specific mutation of the Zic2 locus by microinjection of TALEN mRNA in mouse CD1, C3H and C57BL/6J oocytes. PLoS One 8: e60216
139. Kato T, Miyata K, Sonobe M, Yamashita S, Tamano M, Miura K, Kanai Y, Miyamoto S, Sakuma T, Yamamoto T, Inui M, Kikusui T, Asahara H, Takada S (2013) Production of Sry knockout mouse using TALEN via oocyte injection. Sci Rep 3:3136
140. Qiu Z, Liu M, Chen Z, Shao Y, Pan H, Wei G, Yu C, Zhang L, Li X, Wang P, Fan HY, Du B, Liu B, Liu M, Li D (2013) High-efficiency and heritable gene targeting in mouse by transcription activator-like effector nucleases. Nucleic Acids Res 41:e120
141. Sung YH, Baek IJ, Kim DH, Jeon J, Lee J, Lee K, Jeong D, Kim JS, Lee HW (2013) Knockout mice created by TALEN-mediated gene targeting. Nat Biotechnol 31:23–24
142. Takada S, Sato T, Ito Y, Yamashita S, Kato T, Kawasumi M, Kanai-Azuma M, Igarashi A, Kato T, Tamano M, Asahara H (2013) Targeted gene deletion of miRNAs in mice by TALEN system. PLoS One 8:e76004
143. Wefers B, Meyer M, Ortiz O, Hrabé de Angelis M, Hansen J, Wurst W, Kühn R (2013) Direct production of mouse disease models by embryo microinjection of TALENs and oligodeoxynucleotides. Proc Natl Acad Sci U S A 110:3782–3787

144. Cermak T, Doyle EL, Christian M, Wang L, Zhang Y, Schmidt C, Baller JA, Somia NV, Bogdanove AJ, Voytas DF (2011) Efficient design and assembly of custom TALEN and other TAL effector-based constructs for DNA targeting. Nucleic Acids Res 39:e827
145. Chandrasegaran S, Carroll D (2016) Origins of programmable nucleases for genome engineering. J Mol Biol 428(5 Pt B):963–989
146. Cong L, Ran FA, Cox D, Lin S, Barretto R, Habib N, Hsu PD, Wu X, Jiang W, Marraffini LA, Zhang F (2013) Multiplex genome engineering using CRISPR/Cas systems. Science 339:819–823
147. Jinek M, East A, Cheng A, Lin S, Ma E, Doudna J (2013) RNA-programmed genome editing in human cells. elife 2:e00471
148. Mali P, Yang L, Esvelt KM, Aach J, Guell M, DiCarlo JE, Norville JE, Church GM (2013) RNA-guided human genome engineering via Cas9. Science 339:823–826
149. Mashiko D, Fujihara Y, Satouh Y, Miyata H, Isotani A, Ikawa M (2013) Generation of mutant mice by pronuclear injection of circular plasmid expressing Cas9 and single guided RNA. Sci Rep 3:3355
150. Watts SW, Darios ES, Mullick AE, Garver H, Saunders TL, Hughes ED, Filipiak WE, Zeidler MG, McMullen N, Sinal CJ, Kumar RK, Ferland DJ, Fink GD (2018) The chemerin knockout rat reveals chemerin dependence in female, but not male, experimental hypertension. FASEB J 32:6596–6614
151. Xu J, Zhang L, Xie M, Li Y, Huang P, Saunders TL, Fox DA, Rosenquist R, Lin F (2018) Role of complement in a rat model of paclitaxel-induced peripheral neuropathy. J Immunol 200:4094–4101
152. Allan CM, Heizer PJ, Tu Y, Sandoval NP, Jung RS, Morales JE, Sajti E, Troutman TD, Saunders TL, Cusanovich DA, Beigneux AP, Romanoski CE, Fong LG, Young SG (2018) An upstream enhancer regulates Gpihbp1 expression in a tissue-specific manner. J Lipid Res 60(4):869–879
153. Birling MC, Schaeffer L, André P, Lindner L, Maréchal D, Ayadi A, Sorg T, Pavlovic G, Hérault Y (2017) Efficient and rapid generation of large genomic variants in rats and mice using CRISMERE. Sci Rep 7:43331
154. Boroviak K, Doe B, Banerjee R, Yang F, Bradley A (2016) Chromosome engineering in zygotes with CRISPR/Cas9. Genesis 54:78–85
155. Cheng X, Waghulde H, Mell B, Morgan EE, Pruett-Miller SM, Joe B (2017) Positional cloning of quantitative trait nucleotides for blood pressure and cardiac QT-interval by targeted CRISPR/Cas9 editing of a novel long non-coding RNA. PLoS Genet 13: e1006961
156. Kim JW, Zhang H, Seymen F, Koruyucu M, Hu Y, Kang J, Kim YJ, Ikeda A, Kasimoglu Y, Bayram M, Zhang C, Kawasaki K, Bartlett JD, Saunders TL, Simmer JP, Hu JC (2018) Mutations in RELT cause autosomal recessive amelogenesis imperfecta. Clin Genet 95 (3):375–383. https://doi.org/10.1111/cge.13487
157. Yoshimi K, Kaneko T, Voigt B, Mashimo T (2014) Allele-specific genome editing and correction of disease-associated phenotypes in rats using the CRISPR-Cas platform. Nat Commun 5:4240
158. Yang H, Wang H, Shivalila CS, Cheng AW, Shi L, Jaenisch R (2013) One-step generation of mice carrying reporter and conditional alleles by CRISPR/Cas-mediated genome engineering. Cell 154:1370–1379
159. Mohsen Z, Sim H, Garcia-Galiano D, Han X, Bellefontaine N, Saunders TL, Elias CF (2017) Sexually dimorphic distribution of Prokr2 neurons revealed by the Prokr2-Cre mouse model. Brain Struct Funct 222:4111–4129
160. Mizuhashi K, Ono W, Matsushita Y, Sakagami N, Takahashi A, Saunders TL, Nagasawa T, Kronenberg HM, Ono N (2018) Resting zone of the growth plate houses a unique class of skeletal stem cells. Nature 563:254–258
161. Miura H, Gurumurthy CB, Sato T, Sato M, Ohtsuka M (2015) CRISPR/Cas9-based generation of knockdown mice by intronic insertion of artificial microRNA using longer single-stranded DNA. Sci Rep 5:12799
162. Quadros RM, Miura H, Harms DW, Akatsuka H, Sato T, Aida T, Redder R, Richardson GP, Inagaki Y, Sakai D, Buckley SM, Seshacharyulu P, Batra SK, Behlke MA, Zeiner SA, Jacobi AM, Izu Y, Thoreson WB, Urness LD, Mansour SL, Ohtsuka M, Gurumurthy CB (2017) Easi-CRISPR: a robust method for one-step generation of mice carrying conditional and insertion alleles using long ssDNA donors and CRISPR ribonucleoproteins. Genome Biol 18:92
163. Yoshimi K, Kunihiro Y, Kaneko T, Nagahora H, Voigt B, Mashimo T (2016) ssODN-mediated knock-in with CRISPR-Cas for large genomic regions in zygotes. Nat Commun 7:10431
164. Gu B, Posfai E, Rossant J (2018) Efficient generation of targeted large insertions by

microinjection into two-cell-stage mouse embryos. Nat Biotechnol 36:632–637

165. Paix A, Folkmann A, Goldman DH, Kulaga H, Grzelak MJ, Rasoloson D, Paidemarry S, Green R, Reed RR, Seydoux G (2017) Precision genome editing using synthesis-dependent repair of Cas9-induced DNA breaks. Proc Natl Acad Sci U S A 114: E10745–E10754
166. Yao X, Zhang M, Wang X, Ying W, Hu X, Dai P, Meng F, Shi L, Sun Y, Yao N, Zhong W, Li Y, Wu K, Li W, Chen ZJ, Yang H (2018) Tild-CRISPR allows for efficient and precise gene knockin in mouse and human cells. Dev Cell 45:526–536
167. Codner GF, Mianné J, Caulder A, Loeffler J, Fell R, King R, Allan AJ, Mackenzie M, Pike FJ, McCabe CV, Christou S, Joynson S, Hutchison M, Stewart ME, Kumar S, Simon MM, Agius L, Anstee QM, Volynski KE, Kullmann DM, Wells S, Teboul L (2018) Application of long single-stranded DNA donors in genome editing: generation and validation of mouse mutants. BMC Biol 16:70
168. Gurumurthy C, Quadros R, Adams J Jr, Alcaide P, Ayabe S, Ballard J, Batra SK, Beauchamp M-C, Becker KA, Bernas G, Brough D, Carrillo-Salinas F, Dawson R, DeMambro V, D'Hont J, Dibb K, Eudy JD, Gan L, Gao J, Gonzales A, Guntur A, Guo H, Harms DW, Harrington A, Hentges KE, Humphreys N, Imai S, Ishii H, Iwama M, Jonasch E, Karolak M, Keavney B, Khin N-C, Masamitsu Konno M, Kotani Y, Kunihiro Y, Lakshmanan I, Larochelle C, Lawrence CB, Li L, Lindner V, Liu X-D, Lopez-Castejon G, Loudon A, Lowe J, Jerome-Majeweska L, Matsusaka T, Miura H, Miyasaka Y, Morpurgo B, Motyl K, Nabeshima Y-I, Nakade K, Nakashiba T, Nakashima K, Obata Y, Ogiwara S, Ouellet M, Oxburgh L, Piltz S, Pinz I, Ponnusamy MP, Ray D, Redder RJ, Rosen CJ, Ross N, Ruhe MT, Ryzhova L, Salvador AM, Sedlacek R, Sharma K, Smith C, Staes K, Starrs L, Sugiyama F, Takahashi S, Tanaka T, Trafford A, Uno Y, Vanhoutte L, Vanrockeghem F, Willis BJ, Wright CS, Yamauchi Y, Yi X, Yoshimi K, Zhang X, Zhang Y, Ohtsuka M, Das S, Garry DJ, Hochepied T, Thomas P, Parker-Thornburg J, Adamson AD, Yoshiki A, Schmouth J-F, Golovko A, Thompson WR, Lloyd KC, Wood JA, Cowan M, Mashimo T, Mizuno S, Zhu H, Kasparek P, Liaw L, Miano JM, Burgio G (2018) Re-evaluating one-step generation of mice carrying conditional alleles by CRISPR-Cas9-mediated genome editing technology. BioRxiv. https://doi.org/10.1101/393231
169. Lanza DG, Gaspero A, Lorenzo I, Liao L, Zheng P, Wang Y, Deng Y, Cheng C, Zhang C, Seavitt JR, DeMayo FJ, Xu J, Dickinson ME, Beaudet AL, Heaney JD (2018) Comparative analysis of single-stranded DNA donors to generate conditional null mouse alleles. BMC Biol 16:69
170. Miyasaka Y, Uno Y, Yoshimi K, Kunihiro Y, Yoshimura T, Tanaka T, Ishikubo H, Hiraoka Y, Takemoto N, Tanaka T, Ooguchi Y, Skehel P, Aida T, Takeda J, Mashimo T (2018) CLICK: one-step generation of conditional knockout mice. BMC Genomics 19:318
171. Baker O, Tsurkan S, Fu J, Klink B, Rump A, Obst M, Kranz A, Schröck E, Anastassiadis K, Stewart AF (2017) The contribution of homology arms to nuclease-assisted genome engineering. Nucleic Acids Res 45:8105–8115
172. Jung CJ, Zhang J, Trenchard E, Lloyd KC, West DB, Rosen B, de Jong PJ (2017) Efficient gene targeting in mouse zygotes mediated by CRISPR/Cas9-protein. Transgenic Res 26:263–277
173. Hasty P, Abuin A, Bradley A (2000) Gene targeting, principles, and practice in mammalian cells. In: Joyner AL (ed) Gene targeting: a practical approach. Oxford University Press, Oxford, pp 1–35
174. Müller U (1999) Ten years of gene targeting: targeted mouse mutants, from vector design to phenotype analysis. Mech Dev 82:3–21
175. Saunders TL (2011) Gene targeting vector design for embryonic stem cell modifications. In: Pease S, Saunders TL (eds) Advanced protocols for animal transgenesis: an ISTT manual. Springer-Verlag, Berlin, pp 57–79
176. Anderson KR, Haeussler M, Watanabe C, Janakiraman V, Lund J, Modrusan Z, Stinson J, Bei Q, Buechler A, Yu C, Thamminana SR, Tam L, Sowick MA, Alcantar T, O'Neil N, Li J, Ta L, Lima L, Roose-Girma-M, Rairdan X, Durinck S, Warming S (2018) CRISPR off-target analysis in genetically engineered rats and mice. Nat Methods 15:512–514
177. Kosicki M, Tomberg K, Bradley A (2018) Repair of double-strand breaks induced by CRISPR-Cas9 leads to large deletions and complex rearrangements. Nat Biotechnol 36:765–771
178. Haeussler M, Schönig K, Eckert H, Eschstruth A, Mianné J, Renaud JB, Schneider-Maunoury S, Shkumatava A,

Teboul L, Kent J, Joly JS, Concordet JP (2016) Evaluation of off-target and on-target scoring algorithms and integration into the guide RNA selection tool CRISPOR. Genome Biol 17:148

179. Slaymaker IM, Gao L, Zetsche B, Scott DA, Yan WX, Zhang F (2016) Rationally engineered Cas9 nucleases with improved specificity. Science 351:84–88
180. Kleinstiver BP, Pattanayak V, Prew MS, Tsai SQ, Nguyen NT, Zheng Z, Joung JK (2016) High-fidelity CRISPR-Cas9 nucleases with no detectable genome-wide off-target effects. Nature 529:490–495
181. Vakulskas CA, Dever DP, Rettig GR, Turk R, Jacobi AM, Collingwood MA, Bode NM, McNeill MS, Yan S, Camarena J, Lee CM, Park SH, Wiebking V, Bak RO, Gomez-Ospina N, Pavel-Dinu M, Sun W, Bao G, Porteus MH, Behlke MA (2018) A high-fidelity Cas9 mutant delivered as a ribonucleoprotein complex enables efficient gene editing in human hematopoietic stem and progenitor cells. Nat Med 24:1216–1224
182. Iyer V, Boroviak K, Thomas M, Doe B, Riva L, Ryder E, Adams DJ (2018) No unexpected CRISPR-Cas9 off-target activity revealed by trio sequencing of gene-edited mice. PLoS Genet 14:e1007503
183. Montoliu L, Whitelaw CBA (2018) Unexpected mutations were expected and unrelated to CRISPR-Cas9 activity. Transgenic Res 27:315–319
184. Kumar RA, Chan KL, Wong AH, Little KQ, Rajcan-Separovic E, Abrahams BS, Simpson EM (2004) Unexpected embryonic stem (ES) cell mutations represent a concern in gene targeting: lessons from "fierce" mice. Genesis 38:51–57
185. Westrick RJ, Mohlke KL, Korepta LM, Yang AY, Zhu G, Manning SL, Winn ME, Dougherty KM, Ginsburg D (2010) Spontaneous Irs1 passenger mutation linked to a gene-targeted SerpinB2 allele. Proc Natl Acad Sci U S A 107:16904–16909
186. Khin NC, Lowe JL, Jensen LM, Burgio G (2017) No evidence for genome editing in mouse zygotes and HEK293T human cell line using the DNA-guided Natronobacterium gregoryi Argonaute (NgAgo). PLoS One 12:e0178768
187. Watkins-Chow DE, Varshney GK, Garrett LJ, Chen Z, Jimenez EA, Rivas C, Bishop KS, Sood R, Harper UL, Pavan WJ, Burgess SM (2017) Highly efficient Cpf1-mediated gene targeting in mice following high concentration pronuclear injection. G3 (Bethesda) 7:719–722
188. Zetsche B, Gootenberg JS, Abudayyeh OO, Slaymaker IM, Makarova KS, Essletzbichler P, Volz SE, Joung J, van der Oost J, Regev A, Koonin EV, Zhang F (2015) Cpf1 is a single RNA-guided endonuclease of a class 2 CRISPR-Cas system. Cell 163:759–771
189. Mann JR, McMahon AP (1993) Factors influencing frequency production of transgenic mice. Methods Enzymol 225:771–781
190. Page RL, Canseco RS, Russell CG, Johnson JL, Velander WH, Gwazdauskas FC (1995) Transgene detection during early murine embryonic development after pronuclear microinjection. Transgenic Res 4:12–17

Chapter 2

Pronuclear Microinjection of One-Cell Embryos

Melissa A. Larson

Abstract

Generating a transgenic or gene-modified mouse requires the introduction of exogenous reagents into an early-stage embryo. The mouse one-cell embryo or zygote possesses two pronuclei, representing the genetic contribution of the sperm and oocyte. Traditional transgenic mice are generated by injecting a DNA solution containing a purified transgene construct into the male pronucleus, generally the larger of the two pronuclei. Similarly, gene-editing reagents such as ZFNs, TALENs, and CRISPR RNAs are introduced into zygotes in the same manner, making this technique applicable to a wide variety of projects. This chapter presents the procedures for pronuclear microinjection.

Key words Transgenic, CRISPR, Microinjection, Zygotes, In vivo, Gene editing

1 Introduction

The ability to introduce foreign DNA into a pronucleus of an early embryo and thereby create a transgenic mouse has been a possibility for almost 40 years [1, 2]. In that time, this technique has enjoyed widespread use in transgenic facilities and laboratories throughout the world. Thousands of mice have been generated, resulting in thousands of research articles that have been published employing this technique. A simple PubMed search for "transgenic mice" yields almost 25,000 results, while a Google search results in some 17,000,000 hits. Even with the advent of new genome-editing tools, such as zinc finger nucleases (ZFN), transcription activator-like effector nucleases (TALEN), and CRISPR (clustered regularly interspaced short palindromic repeats), pronuclear micro-injection has remained a valuable method for introduction of these reagents into the embryo for generation of gene-modified mice.

Despite the frequency with which this procedure is employed, injecting a pronucleus of a one-cell mouse embryo is not trivial. The injectionist must be comfortable with handling embryos, with working on a microscope, and primarily with operating micromanipulators. These skilled techniques require the injectionist to

Melissa A. Larson (ed.), *Transgenic Mouse: Methods and Protocols*, Methods in Molecular Biology, vol. 2066, https://doi.org/10.1007/978-1-4939-9837-1_2,

manipulate fine glass tools in three dimensions, as well as providing suction to immobilize an embryo and injection of the desired reagent once the pronuclear membrane has been penetrated. In addition, the equipment to perform pronuclear microinjection is specialized and costly, leading many institutions to establish transgenic core facilities that generate gene-modified mice as a service to research faculty. As such, the transgenic field has recently moved to the generation of CRISPR mice by electroporating reagents into the embryo, although there are limitations of this route of administration.

Because this technique is so fundamental to the transgenic field, the procedures for pronuclear microinjection will be detailed in this chapter.

2 Materials

1. Ten embryo donor female mice of the desired strain (e.g., C57BL/6 or FVB) (*see* **Note 1**).
2. Ten stud male mice of the desired strain, housed individually.
3. Pregnant Mares' Serum Gonadotropin (PMSG), reconstituted to 50 IU/ml in sterile PBS. Aliquot and store at −20 °C. Avoid repeated freezing and thawing.
4. Human Chorionic Gonadotropin (hCG), reconstituted to 50 IU/ml in sterile PBS. Aliquot and store at −20 °C until use. Avoid repeated freezing and thawing.
5. 1-cc insulin syringes with attached 28-gauge needles.
6. M2 HEPES-buffered embryo handling medium, available commercially.
7. KSOM bicarbonate-buffered embryo culture medium, available commercially.
8. Light mineral oil or equivalent.
9. Hyaluronidase, reconstituted to 10 mg/ml in sterile PBS. Make 100 μl aliquots and store at −20 °C until use.
10. Sterile plastic petri dishes, 35 mm and 60 mm.
11. Surgical instruments, including watchmakers' forceps (four pairs), 4″ straight scissors, microscissors.
12. Glass transfer capillary and pipetting apparatus.
13. Stereomicroscope with transmitted light, 8–80× magnification.
14. Injection assembly, including an inverted microscope with 40–400× magnification and Hoffman Modulation Contrast optics (*see* **Note 2**), a pair of manipulators, a micrometer

(CellTram Air or equivalent), and an injector (FemtoJet or equivalent).

15. Fire-polished glass holding pipettes, pulled on a needle puller and polished on a microforge. Alternatively, holding pipettes are available commercially.
16. Glass injection needles with internal filament, pulled on a needle puller or available commercially.
17. Incubator at 37 °C, 6% CO_2, 5% O_2.

3 Methods

3.1 Collection of Presumptive Zygotes

1. Administer 5 IU (0.1 cc) PMSG by intraperitoneal injection to donor females 3 days prior to scheduled pronuclear injection between 12:00 and 2:00 pm.
2. Forty-six to forty-eight hours later, administer 5 IU (0.1 cc) human Chorionic Gonadotropin (hCG) and mate females to individually-housed stud males.
3. Check copulation plugs the next morning.
4. Prepare one 35 mm petri dish with 3 ml M2, one dish with 3 ml M2 plus 100 μl aliquot of 10 mg/ml hyaluronidase (final concentration ~300 μg/ml), one dish with three 100 μl drops of M2 overlaid with oil.
5. Euthanize females according to your IACUC-approved Animal Care and Use Protocol.
6. Lay females on their back and spray abdomen with 70% ETOH.
7. Lift the skin over the abdomen and make a transverse incision with the straight scissors. Pull the skin apart and lift the skin toward the head of each mouse.
8. With clean instruments, lift the body wall of one mouse and cut the tissue open to expose the intestines.
9. Locate the reproductive tract lying along the back wall. The ovaries will lie beneath the kidneys.
10. Grasp one of the uterine horns with forceps and cut away the connective tissue. Pull ever-so-gently to separate the oviduct from the ovary and cut. Slide the forceps up to the bottom of the oviduct and cut beneath the scissors to separate the oviduct from the uterine horn. Place the dissected oviduct in the dish of M2 (without hyaluronidase).
11. Repeat the procedure for the contralateral oviduct in the same female, then repeat **steps 8–11** for all donor females until all 20 oviducts have been collected.
12. With clean forceps, pick up an oviduct and transfer to the dish of M2 with hyaluronidase.

13. With a pair of watchmakers' forceps, pin the oviduct to the bottom of the dish. With a second pair of forceps, tear open the ampulla and release the cumulus mass containing the embryos (*see* **Note 3**).
14. Allow embryos to drop out of the cumulus cells. Pick up clean embryos with a transfer pipette and your pipetting apparatus (mouth pipette) and wash through two drops of M2 (without hyaluronidase). Hold in a drop of M2 overlaid with oil, or place in a drop of KSOM and hold in the incubator at 37 °C, 6% CO_2, 5% O_2.

3.2 Pronuclear Microinjection

1. Prepare an injection dish by placing a 100 μl drop of M2 in the middle of a low-profile 60-mm petri dish. Overlay with oil.
2. Load presumptive zygotes into the drop of M2, slightly South of the center.
3. Move the dish to the stage of the inverted microscope. Lower the holding pipette into the drop of M2 and adjust the height with the micromanipulators until the holding pipette is almost touching the bottom, but it does not scrape the bottom.
4. Fill an injection needle with your reagent to be injected. Injection capillaries are available for purchase that possess an internal filament; these capillaries can be pulled on a pipette puller and then filled by capillary action through back-loading. Simply insert the blunt end of the injection needle into approximately 25 μl of the solution of reagents to be injected and allow to fill.
5. Once the solution passes the neck of the injection needle, remove from the solution and insert into the capillary holder on the micromanipulator. Lower the holding pipette into the drop of M2 while observing at 40× through the binoculars of the microscope. Stop when the injection needle is in focus with the holding pipette.
6. Increase magnification to 200× (or 400× if more comfortable).
7. Prior to injection, it may be necessary to open the tip of the injection needle by breaking on your holding pipette. If so, focus on the tip of the holding pipette, then bring the tip of the injection needle into the same plane of focus by adjusting the depth up and down. Advance the injection needle to ∗just∗ inside the holding pipette and flick up. This should create a small break that will open the injection needle, yet still retain a small enough diameter to not damage the embryo (*see* **Note 4**).
8. Pick up an embryo while applying gentle suction with the holding pipette. Bring the pronuclei of the one-cell embryo into focus. There should be two pronuclei. If only one pronucleus is visible, do not bother to inject as this embryo may be

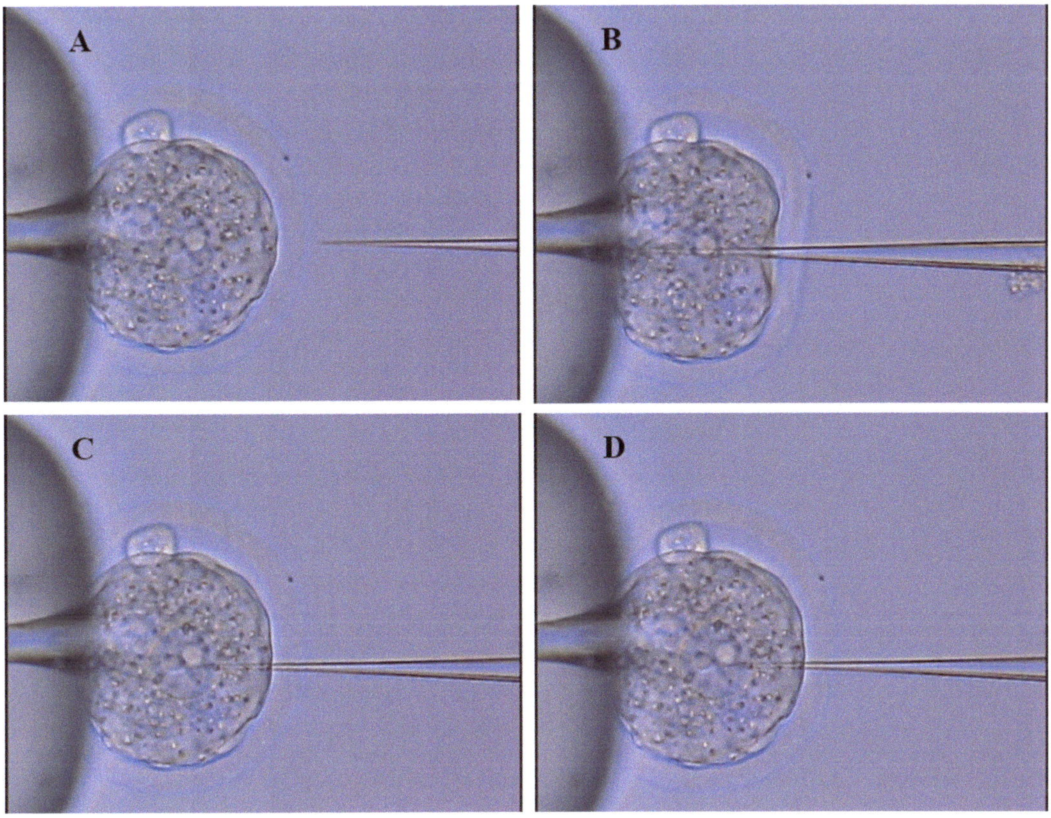

Fig. 1 Pronuclear microinjection of a one-cell embryo. (**a**) An embryo is secured on a fire-polished holding pipette, and two pronuclei are visible. Focus is adjusted until the larger pronucleus appears clear, and the tip of the injection needle is brought into the same plane of focus. (**b**) The injection needle has been inserted beyond the pronucleus to ensure the pronuclear membrane is penetrated. (**c**) The injection needle has been withdrawn to the interior of the pronucleus, and injection has commenced. (**d**) Injection of the reagent solution is visualized by the swelling of the pronucleus (Images courtesy of Illya Bronshteyn, Transgenic and Gene-Targeting Facility, University of Kansas Medical Center, Kansas City, KS, USA)

parthenogenetic. If there are more than two pronuclei, do not inject, as the embryo may be polyspermic and will not develop.

9. Choose to focus on the larger of the two pronuclei. In addition to being easier to inject, injection of the male pronucleus generally results in a higher incidence of integration and activity [3]. Situate the embryo so that the distance from the injection needle to the larger pronucleus is minimized. Focus until the pronuclear membrane is sharp and the interior appears rather clear. Adjust the height of the injection needle until the tip of the needle is focused on the same plane as the pronuclear membrane (Fig. 1).
10. Propel the injection needle forward until it touches the zona pellucida (the thick glycoprotein coat surrounding the embryo) and gives a gentle nudge. This will give you some

indication as to whether you have adjusted the needle to the middle of the embryo in the z plane. If you are too high or too low, you will likely roll the embryo.

11. Once you have established that you are near the middle of the embryo and the tip of the needle is in focus with the pronuclear membrane, give a short, controlled jab and insert the needle into the pronucleus. With the needle in the pronucleus, try to expel some solution by stepping on the pedal of the microinjector (*see* **Note 5**). If no swelling is noted, it may be necessary to actually pass beyond the pronucleus and draw back until the tip of the needle is within the cavity. Try to expel the solution again. Once pronuclear swelling is observed, withdraw the needle from the embryo (Fig. 1).
12. Move the injected embryo to the top of the injection drop, return to the center, pick up another embryo, and repeat the injection process.
13. When all embryos have been microinjected, move the injected embryos into a drop of KSOM for overnight culture in the incubator.
14. Score injected embryos for development to the two-cell stage. Transfer two-cell embryos to recipient females, 20–25 per recipient.

4 Notes

1. The strain of choice for creating a gene-modified mouse is highly variable. Embryos from the inbred FVB strain have clear cytoplasm and large, readily visible pronuclei, making them ideal for microinjection. However, many researchers prefer mice with a C57BL/6 background. Unfortunately, C57BL/6 embryos have a granular cytoplasm and small pronuclei, making it difficult to visualize the pronuclear target for microinjection.
2. Alternatively, the inverted microscope may be equipped with differential interference contrast (DIC) optics, but these optics require use of glass injection dishes or slides.
3. It is actually more like scoring the oviduct near the ampulla (the swollen and clear area of the oviduct). The cumulus mass will be visible in the ampulla, and you need only to nick the oviduct to create an opening; the cumulus mass will ooze out on its own.
4. Clearly, the technique of breaking open a needle is a skill learned through trial and error (and many hours on the microscope). If the needle is too small, no solution will be injected; too large, and you will lyse the embryo.

5. The pressure at which you microinject will be determined by the injection system you are using, and this will have to be established empirically. Remember to start low (e.g., 30 psi) and increase the pressure gradually if need be. It may also be that the tip of the injection needle needs to be broken open. It is also helpful to have a low constant pressure set on the microinjector (15–20 psi) to prevent M2 medium from being aspirated into the needle.

References

1. Gordon JW, Scangos GA, Plotkin DJ, Barbosa JA, Ruddle FH (1980) Genetic transformation of mouse embryos by microinjection of purified DNA. Proc Natl Acad Sci U S A 77:7380–7384
2. Gordon JW, Ruddle FH (1981) Integration and stable germ line transmission of genes injected into mouse pronuclei. Science 214 (4526):1244–1246
3. Brinster RL, Chen HY, Trumbauer ME, Yagle MK, Palmiter RD (1985) Factors affecting the efficiency of introducing foreign DNA into mice by microinjecting eggs. Proc Natl Acad Sci U S A 82:4438–4442

Chapter 3

Integrase-Mediated Targeted Transgenics Through Pronuclear Microinjection

Ruby Yanru Chen-Tsai

Abstract

Transgenic technology allows a gene of interest to be introduced into the genome of a laboratory animal and provides an extremely powerful tool to dissect the molecular mechanisms of disease. Transgenic mouse models made by microinjection of DNA into zygotic pronuclei, in particular, have been widely used by the genetics community for over 35 years. However, up till 5 years ago, it remained a rather crude approach: injected sequences randomly insert in multiple copies as concatemers, and they can be mutagenic and have variable, ectopic, or silenced expression depending on the site of integration, a phenomenon called position effects. As a result, multiple lines are required in order to confirm appropriate transgene expression. This can be partially overcome by flanking transgenes with insulator sequences to protect the transgene from influence of surrounding regulatory elements. Large (<300 kb) BAC-based transgenic vectors have also been shown to be more resistant to position effects. However, animals carrying extra copies of fairly large regions of the genome could have unpredictable phenotypes.

These problems can be overcome by targeting the transgene to a specific chromosomal locus via homologous recombination in embryonic stem (ES) cells. However, this method is significantly more laborious and time consuming, as it involves creation of modified ES cells and mouse chimeras, as well as eventual germline transmission of the transgene.

Here, I describe an integrase-based approach, trademarked as "TARGATT™" (target *attP*), to produce site-specific transgenic mice via pronuclear microinjection, whereby an intact single-copy transgene can be inserted into predetermined chromosomal loci with high efficiency (up to 40%), and faithfully transmitted through generations. This system allows high-level global transgene expression or tissue-specific expression depending on the promoter used, or inducible expression such as induced by tetracycline or doxycycline. Using this approach, site-specific transgenic mice can be generated as fast as in 3 months. The technique presented here greatly facilitates murine transgenesis and precise structure/function dissection of mammalian gene function and regulation in vivo.

Key words Pronuclear microinjection, PhiC31 integrase, TARGATT™, Site-specific transgenesis, *Rosa26* knockin, *H11* locus

1 Introduction

The demonstration that sequence-specific recombinases, such as integrases derived from lower organisms, could function in mammalian cells provided more tools for mammalian genome

Melissa A. Larson (ed.), *Transgenic Mouse: Methods and Protocols*, Methods in Molecular Biology, vol. 2066, https://doi.org/10.1007/978-1-4939-9837-1_3, © Springer Science+Business Media, LLC, part of Springer Nature 2020

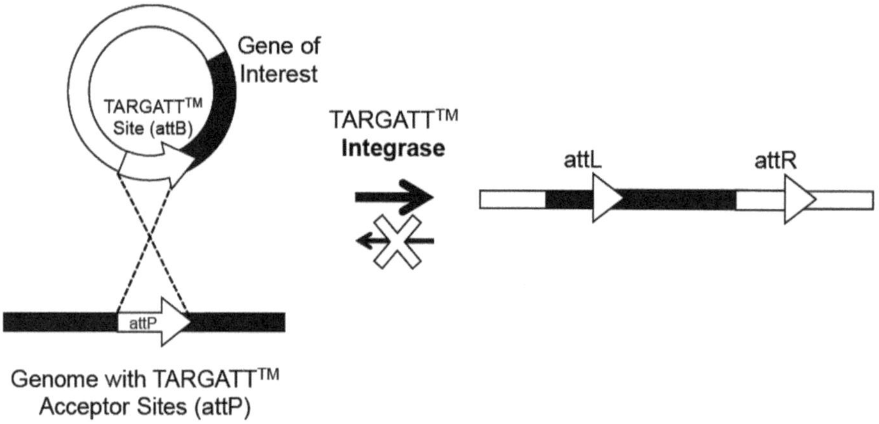

Fig. 1 Site-specific transgene integration mediated by phage integrases. Integrases such as phiC31 integrase mediate sequence-specific recombination between two largely different sites, *attP* and *attB*. The sequences of the *attP* and *attB* are unrelated except for the three nucleotides in the middle where the crossover happens. Recombination between *attP* and *attB* results in two new hybrid sites *attL* and *attR*, which can no longer be recognized by integrase. So integrase is unidirectional, catalyzing only integration but no reverse excision reaction

manipulation [1–3]. Integrases catalyze recombination between appropriate *attB* and *attP* sites [1, 4], resulting in an integration of the DNA sequence flanking the *attB* site (Fig. 1). At the same time, recombination between *attP* and *attB* results in two new hybrid sites, *attL* and *attR*. These two new sites can no longer be recognized by integrases. This irreversible feature, along with the lack of a corresponding excisionase enzyme, makes the recombination reaction unidirectional, ensuring that constructs integrated into the genome do not act as substrates for the reverse reaction. The result is an improvement in integration efficiency compared with random integration. The integrase recognition sites, *attP* or *attB*, do not naturally exist in mammalian genome. Instead, sites with similar sequences so-called pseudo-*attP* sites can also be recognized by integrase such as PhiC31 integrase from a Streptomyces phage. The efficiency of recombination between an *attB* and a pseudo-*attP* site is low compared to its wild-type cognate *attP* site. PhiC31 integrase has been used to catalyze integration of circular DNA into pseudo-*attP* sites in the mouse genome for gene therapy [5], for low efficiency transgenesis in mice [6], and for integrating DNA in bovine cells [3].

The TARGATT™ transgenic technology has the mouse genome pre-engineered with wild-type *attP* sites in a safe genomic harbor locus and allows irreversible and site-specific transgene integration by direct pronuclear microinjection into the mouse zygotes (Fig. 2). TARGATT™ mice containing the *attP* sites were originally generated by inserting the *attP* sites in one of the two precharacterized, intergenic, and transcriptionally active loci, *H11* [7]

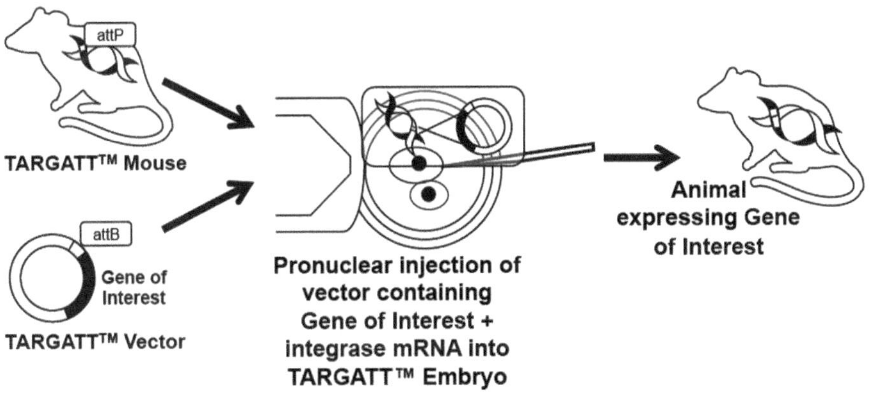

Fig. 2 Generation of site-specific transgenic mice using TARGATT™. One-cell fertilized embryos (zygotes) are harvested from TARGATT mice that contain the *attP* site at either *Rosa26* or *H11* locus. TARGATT donor vector containing the *attB* site and the gene of interest are microinjected into the pronucleus together with integrase mRNA. Injected embryos are implanted into the oviduct of pseudopregnant females to carry the embryos to term

or *Rosa26*, using homologous recombination in mouse embryonic stem cells [8]. Between the two safe harbor loci, *Rosa26* has been used widely as a site for gene knockin since 1999 [9]. The *H11* locus is located in between two highly expressed genes and in an intergenic region on mouse chromosome 11. The *H11* locus was originally described by Hippenmeyer et al. [7]. Further in vitro and in vivo experiments showed that this locus supports robust, high-level gene expression.

For generating site-specific transgenic mice, TARGATT™ mice containing the *attP* docking sites at the *H11* or *Rosa26* locus are used as zygote donors. A mixture of a transgene vector containing a compatible site, *attB*, and in vitro transcribed phiC31 mRNA is microinjected into a pronucleus of each zygote. Figure 2 illustrates how this integrase-based TARGATT™ method works for generating site-specific transgenic mice. PhiC31 catalyzes recombination between *attP* and *attB*, resulting in integration of the gene of interest (*see* **Note 1**).

2 Materials

2.1 TARGATT™ DNA Plasmid Preparation

1. Several selections of the TARGATT™ plasmid can be purchased from Applied StemCell, Inc. (Milpitas, CA). Each plasmid contains a unique combination of a promoter, loxP-stop-loxP (LSL) cassette, multiple cloning site (MCS), and a polyA sequence to meet each project needs. A list of the plasmids includes the following:

    (a) TARGATT™ 2 (pCAG-MCS-polyA).

   (b) TARGATT™ 3 (no promoter +MSC).
   (c) TARGATT™ 6.1 (pCAG-LSL-MCS-polyA).
   (d) TARGATT™ 7 (pPGK-MCS-polyA).
   (e) TARGATT™ 8 (pCA-MCS-polyA).
   (f) TARGATT™ 9.1 (pCA-LSL-MCS-polyA).
2. Use an endotoxin-free plasmid preparation kit to purify plasmid DNA. Phenol/chloroform is used at the final purification step.

2.2 Targeted Transgenic Mice

1. Ten TARGATT™ mice are needed for each round of pronuclear microinjection experiment (*see* **Note 2**).
   (a) Strain code 537: TARGATT™ mice with *attP* sites at the *H11* locus, FVB strain from Charles River Laboratories (CRL).
   (b) Strain code 549: TARGATT™ mice with *attP* sites at the *H11* locus, C57BL6 strain from CRL.
   (c) AST-0013: Cryopreserved embryos with *attP* sites at the *Rosa26* locus, C57BL6 strain from Applied StemCell, Inc. (ASC).
   (d) AST-0004: Cryopreserved embryos with *attP* site at the *Rosa26* locus, FVB strain from ASC.
2. Ten vasectomized male mice and 20 CD1 mature female mice can be purchased from CRL or other animal vendors. After mating, the females are used as pseudopregnant females for embryo implantation.
3. Transgenic kits can be purchased from Applied StemCell, Inc. Each kit contains the following:
   (a) TARGATT™ integrase mRNA (transgene integration tested) (200 ng/μl).
   (b) Control mRNA (50 ng/μl).
   (c) Embryo qualified microinjection TE (MiTE) buffer.
4. Dissection tools, embryo culture media, a microinjection setup, and stereo microscopes used for routine pronuclear microinjection are needed. Supplies such as pipettes, pipette puller, sutures, wound clips, and heating pads routinely used for oviduct embryo transfer are needed.

2.3 Genotyping Founder Animals

The genotyping kits for detecting site-specific transgene insertion can be purchased from Applied StemCell, Inc. Each kit contains the following:

(a) Tail tissue lysis buffer.

(b) Proteinase K (20 mg/ml).

(c) PCR primer sets for detecting site-specific insertion at 5′ and 3′ junctions, including H11P3, Rosa26P3, SSL, and SSR.
(d) Nuclease-free water.
(e) Positive-control DNA.

3 Methods

3.1 Preparation of Plasmid DNA for Microinjection

The TARGATT™ plasmids contain multiple cloning site (MCS) which can be used to insert the transgene (or gene of interest). Common steps are listed as the following (Fig. 3).

1. Cut and purify the insertion DNA fragment from a donor vector or have the DNA fragment synthesized.
2. Linearize TARGATT™ plasmid with restriction enzyme (s) in MCS.
3. Ligate TARGATT™ plasmid backbone with the insertion DNA fragment.
4. Use an endotoxin-free plasmid kit (Qiagen) to extract TARGATT™ DNA insertion vector.
5. Purify DNA with phenol/chloroform extraction twice and use phase-separation gel.
6. Determine plasmid DNA concentration on a gel using a DNA mass standard.
7. Determine plasmid DNA purity using spectrophotometer (A260/A280>1.8).
8. Digest plasmid DNA with restriction enzymes to confirm the integrity of the plasmid.
9. Test for the absence of RNase by incubating 200 ng plasmid DNA with 200 ng control RNA at 37 °C for 1 h, then heat-inactivate at 85–90 °C for 5 min, and store on ice for 2 min.

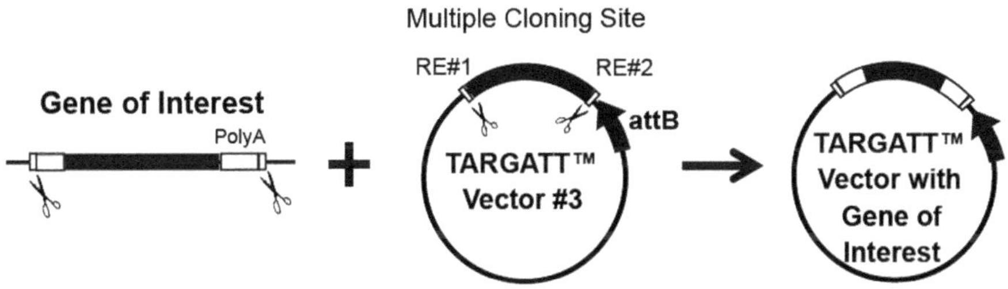

Fig. 3 Construction of TARGATT™ donor vector. The donor vector can usually be made by one-step cloning. Transgene can either be synthesized or cut out from another vector and subcloned into a TARGATT™ vector (available from Applied StemCell, Inc.). There are several versions of TARGATT vectors including choice of promoter and stop cassette for conditional expression, but all containing the *attB* site and multiple cloning site (MCS) for easy subcloning

Run the mixture on a 1% agarose gel at 130 V for 15 min. Gel box should be kept cold during the run. There should be no RNA degradation shown as a smear on a gel.

10. Dilute the DNA with microinjection TE buffer (MiTE) supplied in the TARGATT™ transgenic kit to a final concentration of 6 ng/μl and store it at −80 °C until microinjection.

3.2 Generating Site-Specific Transgenic Founders

1. Mix equal volume (10 μl each) of the plasmid DNA (6 ng/μl) and TARGATT™ integrase mRNA supplied in the TARGATT™ transgenic kit. The mixture should be made fresh on the day of microinjection. Filter the mixture through an RNase-free 0.2 μm syringe filter right before microinjection. The 20 μl of mixture should be sufficient for injecting 200 embryos.
2. The mixture of insertion plasmid DNA and integrase mRNA is microinjected into the pronuclei of one-cell embryos (zygotes) using a constant flow mode of the microinjector, so the mixture is distributed into both the pronuclei and cytoplasm. If the embryos are homozygous for TARGATT *attP* sites, either of the two pronuclei can be injected. But if the embryos are from TARGATT *attP* male bred with wild-type females, the injection should be into the male pronuclei. In the case when it is not easy to tell the male pronucleus from the female one, inject both pronuclei to ensure integrase is present in the *attP*-containing nucleus.
3. Upon completion of microinjection, embryos are implanted into the oviduct of pseudopregnant females at about 20–25 embryos per female to carry the embryos to term.

3.3 Identifying Founders by Genotyping

1. Upon weaning the pups (2–3 weeks after birth), tail snips can be obtained from each pup for genotyping using a TARGATT™ Mouse Genotyping Kit.
2. Cut approximately 2 mm tail snip and place in an Eppendorf tube.
3. Add 200 μl tail tissue lysis buffer and Proteinase K, and lyse at 55 °C for 4 h to overnight.
4. Heat the samples at 90 °C for 5 min to inactivate proteinase K.
5. Proceed to PCR reaction or store at −20 °C.
6. Set up PCR reactions on ice according to instruction of the Taq polymerase manufacturer. Use table below for Qiagen Tag polymerase.

Component	Amount (μl)
DNA—Tail lysis	1
10× PCR buffer	2
dNTPs (10 mM)	0.4
Primer set (10 μM)[a]	1
Taq polymerase (5 U/μl)	0.2
Nuclease-free water	15.4
Total volume	20

[a]Primer sets for genotyping:

(a) Use "Primer Set H11P3" (AST-2005) or "Primer Set Rosa26P3" (AST-2006) to examine whether the mouse has *attP* insertion at the correct locus (H11 or Rosa26).

(b) Use "Primer Set SSL" and "Primer Set SSR" to determine site-specific gene integration.

7. Perform PCR amplification using the following program.

Step	No. of cycles	Temperature (°C)	Time
1	1	94	3 min
2	35	94	30 s
3		58	25 s
4		72	45 s
5	1	72	5 min

8. Run PCR products to analyze.
9. Figure 4 depicts examples of TARGATT™ genotyping scheme on identifying site-specific transgenic founders. Primer sets H11P3.F (forward primer)/H11P3.R (reverse primer) and R26P3.F/R26P3.R are used to detect the *attP*x3 at the *H11* locus or *Rosa26* locus, respectively, before transgene insertion. Here is a list of PCR products based on the genotype at the locus.

(a) 690 bp PCR product for *attP*x3-modified *H11* allele.

(b) 364 bp PCR product for wild-type *H11* allele.

(c) *H11* heterozygotes show both the 690 bp and 364 bp bands.

(d) 663 bp PCR product for *attP*x3-modified *Rosa26* allele.

(e) 278 bp PCR product for wild-type *Rosa26* allele.

(f) *Rosa26* heterozygotes show both the 663 bp and 278 bp bands.

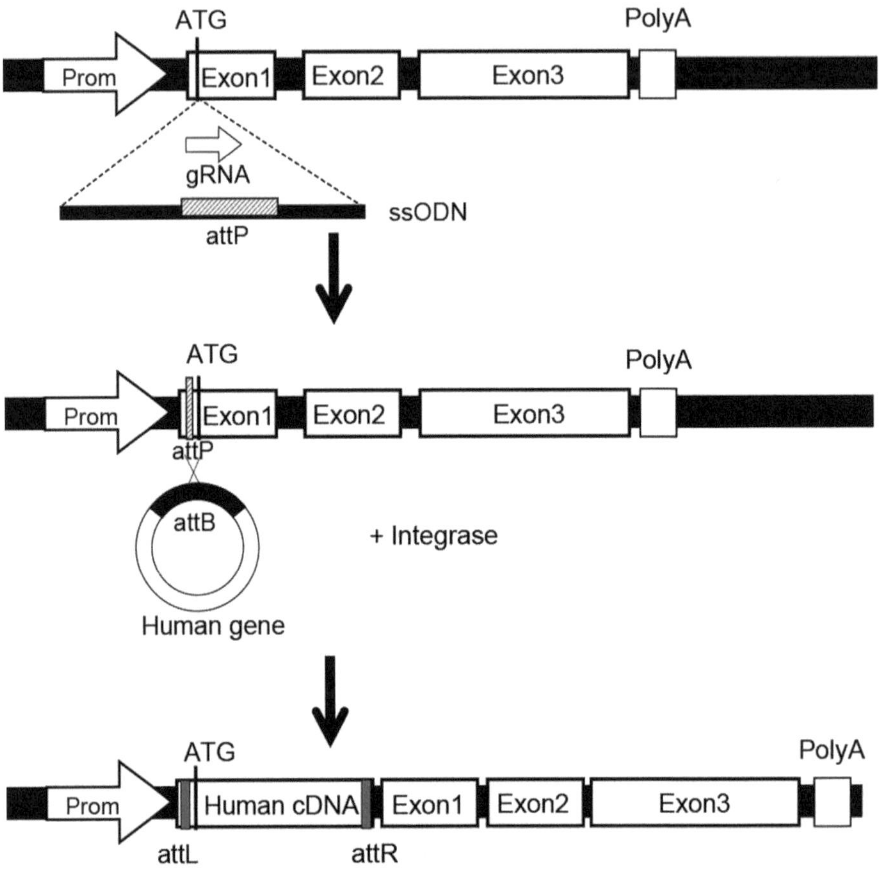

Fig. 4 TARGATT™ genotyping data analysis. The TARGATT™ mice contain three copies of the *attP* site (*attP*x3) in tandem at either the *H11* or *Rosa26* locus. Site-specific recombination between an *attP* site on the genomic DNA and the *attB* site on the TARGATT™ vector results in transgene integration. Depending on which one of the three *attP* sites is recombined with the *attB* site, different sizes of PCR fragments are expected. The example shown here is a result of recombination of the first *attP (attP1)* site with the *attB* site

10. To detect site-specific transgenic founders at both the 5′ and 3′ junctions, two sets of PCR are carried out; one with primer set SSL.F/SSL.R for 5′ and one with SSR.F/SSR.R for 3′ junction fragment (Fig. 4). Anticipated sizes of PCR fragments using primer set SSL are as follows:

    (a) 136 bp PCR product for 5′ insertion at *attP* site 1.

    (b) 206 bp PCR product for 5′ insertion at *attP* site 2.

    Anticipated sizes of PCR fragments using primer set SSR are as follows:

    (c) 225 bp PCR product for insertion at *attP* site 2.

    (d) 55 bp PCR product for insertion at *attP* site 3.

4 Notes

1. The discovery of CRISPR/Cas9 technology [10–12] has provided an unprecedented, revolutionary, and relatively easy method for genome editing both in cells and in animals. The Cas9 protein makes site-specific double-stranded DNA breaks, and then uses the host recombination machinery including nonhomologous end joining (NHEJ) for error-prone repair or homology-directed repair (HDR) for precise DNA insertion. One limitation of CRISPR is that this method does not work in host cells that have inactive DNA recombination pathways such as nondividing or slow-dividing cells. NHEJ is a major pathway which generates indels, resulting in frameshift. This feature is used as a strategy for generating knockout mice. However, HDR happens in a very low rate in a host cell. As a result, DNA insertion rate is low using CRISPR/Cas9 when ssODNs (single-stranded oligodeoxynucleotides <200 bp) are used as donor DNA. When plasmid DNA is used as a donor, knockin efficiency is even lower. We have not been successful in inserting fragments larger than 3 kb using CRISPR/Cas9. In comparison, phiC31 integrase is self-autonomous such that the enzyme itself contains all the functional components for recombination including DNA cutting, strand exchange, and re-ligation and is, therefore, independent of the host cell recombination pathways. As a result, TARGATT™ works in nondividing cells. Another advantage of TARGATT is its higher efficiency in DNA insertion in transgenic mice, with 5–40% live-born animals being positive founders, and DNA fragments up to 22 kb have been successfully inserted. Table 1 is a comparison of TARGATT™ with CRISPR/Cas9 and recommends which technology to use for specific transgenic models. More importantly, the combination of CRISPR and TARGATT can be used to do gene replacement directly in embryos, bypassing the need for an ES cell homologous recombination step. The first step of this method is to knock in an *attP* site (~50–70 bp) at a desired genomic site using CRISPR, followed by a second step to insert a transgene at the *attP* site using TARGATT (Fig. 5). In summary, CRISPR/Cas9 and TARGATT™ are two complementary technologies, and together they provide a set of genetic tools for making diverse genetic modifications in mice.
2. Since the first few papers on using PhiC31 to generate site-specific transgenic mice was published [8, 13, 14], we have successfully generated site-specific transgenic mouse models for over 80 DNA constructs using the TARGATT™ technology. Besides the models described in the publications of which ASC employees are co-authors, examples of TARGATT models

Table 1
Choosing the right technologies: CRISPR/Cas9 or TARGATT™

Project purpose	CRISPR/Cas9	TARGATT™
Knockout (KO)	Efficient	
Point mutation	Yes	
Knockin (200 nucleotides-2 kb)	Yes (<2 kb)	
Knockin transgene in a safe genomic locus		Yes (up to 20 kb)
Knockin gene replacement (2-step approach) (Fig. 4)	First step Knockin *attP* site	Second step Insert transgene

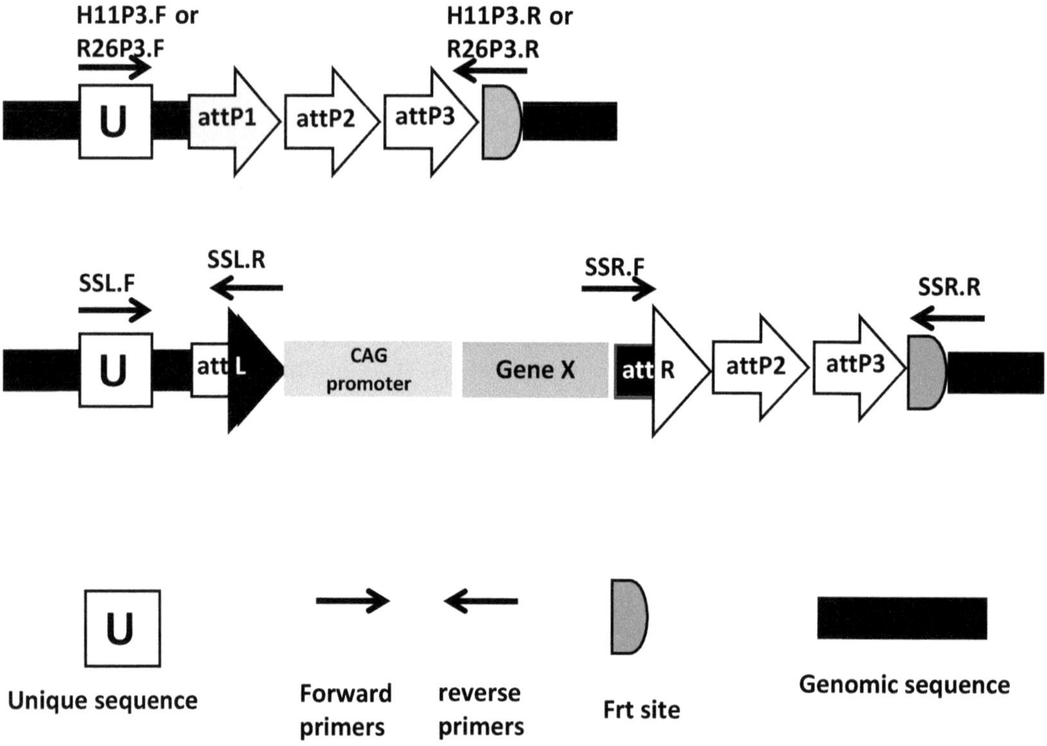

Fig. 5 Making gene replacement (e.g., humanized mouse model) models by pronuclear microinjection using CRISPR/Cas9 and TARGATT™. As a first step, a donor ssODN (single-stranded oligodeoxynucleotide <200 bp) containing an *attP* site (~50–70 bp) is knocked in at a desired genomic locus using CRISPR/Cas9 method. Transgene cassette (up to 20 kb) can then be inserted at the *attP* site using the TARGATT™ method. "Prom" stands for promoter

generated by Applied StemCell, Inc. for its customers are published in reference no. [15–18]. A data summary of 25 projects indicates that many factors contribute to the outcome of the experiments and some of the general guidelines are detailed as the following.

(a) Strain background: The TARGATT™ mice are in two strain backgrounds, C57BL6/J and FVB. There is no statistical difference between the two strains in terms of efficiency.

(b) The efficiency at the *Rosa26* locus appears to be slightly higher than the *H11* locus with an average efficiency of 10% at *Rosa26* versus 8% at *H11*. The overall efficiency varied, ranging from 0.3% to 40%.

(c) Insertion size: There is a general trend that the bigger the insertion is, the lower the efficiency tends to be. With this method, the largest fragment we successfully inserted is 22 kb.

(d) Minicircle DNA versus plasmid DNA: There is no statistical difference in insertion efficiency between using a minicircle vector with one *attB* site compared to a plasmid DNA with two *attB* sites.

Acknowledgments

The author thanks all Applied StemCell, Inc. employees who have contributed to improving the TARGATT™ technologies. Special thanks to Maki Ogawa for helping with the figures.

References

1. Groth AC, Olivares EC, Thyagarajan B, Calos MP (2000) A phage integrase directs efficient site-specific integration in human cells. Proc Natl Acad Sci U S A 97:5995–6000
2. Keravala A, Groth AC, Jarrahian S, Thyagarajan B, Hoyt JJ, Kirby PJ, Calos MP (2006) A diversity of serine phage integrases mediate site-specific recombination in mammalian cells. Mol Gen Genomics 276:135–146
3. Ma QW, Sheng HQ, Yan JB, Cheng S, Huang Y, Chen-Tsai Y, Ren ZR, Huang SZ, Zeng YT (2006) Identification of pseudo attP sites for phage phiC31 integrase in bovine genome. Biochem Biophys Res Commun 345:984–988
4. Thorpe HM, Smith MC (1998) In vitro site-specific integration of bacteriophage DNA catalyzed by a recombinase of the resolvase/invertase family. Proc Natl Acad Sci U S A 95:5505–5510
5. Olivares EC, Hollis RP, Chalberg TW, Meuse L, Kay MA, Calos MP (2002) Site-specific genomic integration produces therapeutic factor IX levels in mice. Nat Biotechnol 20:1124–1128
6. Hollis RP, Stoll SM, Sclimenti CR, Lin J, Chen-Tsai Y, Calos MP (2003) Phage integrases for the construction and manipulation of transgenic mammals. Reprod Biol Endocrinol 1:79
7. Hippenmeyer S, Youn YH, Moon HM, Miyamichi K, Zong H, Wynshaw-Boris A, Luo L (2010) Genetic mosaic dissection of Lis1 and Ndel1 in neuronal migration. Neuron 68:695–709
8. Tasic B, Hippenmeyer S, Wang C, Gamboa M, Zong H, Chen-Tsai Y, Luo L (2011) Site-specific integrase-mediated transgenesis in mice via pronuclear injection. Proc Natl Acad Sci U S A 108:7902–7907
9. Soriano P (1999) Generalized lacZ expression with the ROSA26 Cre reporter strain. Nat Genet 21:70–71
10. Cong L, Ran FA, Cox D, Lin S, Barretto R, Habib N, Hsu PD, Wu X, Jiang W, Marraffini LA, Zhang F (2013) Multiplex genome engineering using CRISPR/Cas systems. Science 339:819–823

11. Mali P, Yang L, Esvelt KM, Aach J, Guell M, DiCarlo JE, Norville JE, Church GM (2013) RNA-guided human genome engineering via Cas9. Science 339:823–826
12. Wang H, Yang H, Shivalila CS, Cheng AW, Shi L, Jaenisch R (2013) One-step generation of mice carrying mutations in multiple genes by CRISPR/Cas-mediated genome engineering. Cell 153(4):910–918
13. Fan X, Petitt M, Gamboa M, Huang M, Dhal S, Druzin ML, Wu JC, Chen-Tsai Y, Nayak NR (2012) Transient, inducible, placenta-specific gene expression in mice. Endocrinology 153:5637–5644
14. Guenther CA, Tasic B, Luo L, Bedell MA, Kingsley DM (2014) A molecular basis for classic blond hair color in Europeans. Nat Genet 46(7):748–752
15. Booze ML, Hansen JM, Vitiello PF (2016) A novel mouse model for the identification of thioredoxin-1 protein interactions. Free Radic Biol Med 99:533–543
16. Feng D, Dai S, Liu F, Ohtake Y, Zhou Z, Wang H et al (2016) Cre-inducible human CD59 mediates rapid cell ablation after intermedilysin administration. J Clin Invest 126 (6):2321–2333
17. Sun N, Yun J, Liu J, Malide D, Liu C, Rovira L et al (2015) Measuring in vivo mitophage. Mol Cell 60(4):685–696
18. Devine WP, Wythe JD, George M, Koshiba-Takeuchi K, Bruneau BG (2014) Early patterning and specification of cardiac progenitors in gastrulating mesoderm. elife 3:e03848

Chapter 4

In Vivo Validation of CRISPR Reagents in Preimplantation Mouse Embryos

Melissa A. Larson, Katelin A. Gibson, and Jay L. Vivian

Abstract

The CRISPR/Cas9 system has enjoyed enormous success and has now become the standard method of generating gene-modified mouse models. The tools for predicting the activity of CRISPR reagents in the mouse embryo are currently limited and not particularly accurate in predicting if a given reagent will be active. Given the time and cost of generating genetically modified mice, it is highly desirable to use CRISPR reagents that are known to be active in the mouse embryo. In this chapter, we provide a detailed procedure for empirically testing the activity of CRISPR reagents via electroporation into cultured preimplantation mouse embryos. This platform has proven to be rapid, efficient, and applicable to a variety of mouse strains, and can be used for assessing on- and off-target activity through a variety of molecular assays.

Key words CRISPR, Validation, crRNA, gRNA, Indels, Cas9, Electroporation, NEPA21, Embryo, Zygote

1 Introduction

The clustered regularly interspaced short palindromic repeats/ CRISPR associated protein (CRISPR/Cas9) system has become a central tool for generating genetically modified mouse models [1]. The utility of CRISPR/Cas9 for making specific mutations in the mouse requires reagents that are both highly active and highly specific. Various bioinformatics tools are available to identify guide RNA (gRNA) target sites nearby the genomic region to be altered [2, 3], and putative off-target sites for a given gRNA can also be identified using algorithms to identify similar sequences in the genome. However, the tools for predicting the activity of CRISPR reagents that target a given site [4] are currently limited and not particularly accurate in predicting if a given gRNA will be active. Our empirical data have shown that approximately half of all CRISPR reagents do not show substantial activity in the mouse embryo. Given the time and cost of generating genetically modified mice, it is highly desirable to use CRISPR reagents that are known

Melissa A. Larson (ed.), *Transgenic Mouse: Methods and Protocols*, Methods in Molecular Biology, vol. 2066, https://doi.org/10.1007/978-1-4939-9837-1_4,

to be active in the mouse embryo, and a rapid means of empirical determination of CRISPR reagent activity is warranted.

Testing the activity of a CRISPR reagent has often involved the transfection of the reagent into cell lines [5]. The target site is then assessed for small insertions and deletions generated through non-homologous end joining (NHEJ) as a readout of the qualitative and quantitative activity of the CRISPR reagent. The capacity for CRISPR/Cas9 to bind and introduce double-stranded breaks is highly dependent on the local chromatin structure of a given target. We have observed that many CRISPR reagents that show activity after transfection in a mouse cell line do not necessarily display activity in the early mouse embryo. Given the rather unusual and dynamic chromatin structure of the early embryo, it is perhaps not surprising that CRISPR reagents which show activity in cell lines, may not necessarily be active in the mouse zygote.

In this chapter, we provide a detailed procedure we have developed for introducing CRISPR reagents into cultured preimplantation mouse embryos for assessment of their activity. This efficient protocol provides a readout of CRISPR reagent activity in the context of the mouse embryo, to determine if a reagent is suitable for subsequent use in the production of novel genetically modified mice. This system leverages mouse embryo electroporation [6] to efficiently introduce Cas9 protein complexed with either gRNA or crRNA/tracrRNA as a ribonucleoprotein (RNP) complex into the zygote. These embryos are cultured for 4 days to the blastocyst stage and are then harvested for isolation of genomic DNA for use in a polymerase chain reaction (PCR) to assess CRISPR activity of the target site via assessment of indels generated through NHEJ. This platform has proven to be rapid, efficient, and applicable to a variety of mouse strains. The basic electroporation procedure and PCR analysis are rapid and can be completed in approximately 1 week, in addition to the 2 weeks in advance required to obtain females for superovulation and embryo collection and to synthesize the CRISPR reagents. The procedure is also extensible to a variety of analyses. Although the validation can be assessed using single pass Sanger sequencing to assess activity, other assays to detect NHEJ-mediated indels can be used, such as T7 endonuclease assay, if more quantitative assessment is needed. These PCR products are also highly amenable for use in deep sequencing, which allows for a highly quantitative assessment of indels via NHEJ and also provides a mutation profile of indel size and location [5, 7]. The use of pooled embryos provides sufficient genomic DNA for additional testing, such as examination of off-target activity, if desired. Given that single-stranded oligonucleotide (ssODN) donor DNA can also be introduced by electroporation, this assay can also be extended to assess the activity of homology directed repair, with little modification to the procedure.

2 Materials

2.1 Embryo Collection

1. Ten prepubertal C57BL/6 female mice (C57BL/6 superovulate best between 3 and 4 weeks of age). Other strains can be used, including FVBn females mated to either FVBn or C57Bl/6 males (but *see* **Note 1** regarding use of F1 embryos). Optimal age for superovulation will differ depending upon the strain [8].
2. Ten sexually mature C57BL/6 male mice, housed individually.
3. Pregnant mare's serum gonadotropin (PMSG) reconstituted to 50 IU/ml. Pharmaceutical grade PMSG may be provided by P.G. 600, which is a combination of PMSG and human chorionic gonadotropin (hCG) intended for the superovulation of females. One vial contains 2000 IU of lyophilized PMSG. To reconstitute: Add 40 ml of sterile phosphate-buffered saline (PBS) to a 50 ml conical tube. Withdraw 5 ml of PBS from the conical tube with a syringe and needle and dispense in the vial of P.G. 600. Swirl the solution to allow the PMSG to dissolve, then aspirate the solution and add to the remaining volume of PBS in the conical tube. Mix by capping and inverting the tube several times. Aliquot 1ml to 1.5 ml microcentrifuge tubes and store at -80° until use.
4. Human chorionic gonadotropin reconstituted to 50 IU/ml. This hormone is available in vials of 2500 IU from Sigma. To reconstitute: Add 50 ml sterile PBS to a 50 ml conical tube. Follow directions for reconstitution as described for PMSG—withdraw PBS from the conical tube, transfer to the vial of hCG, swirl, and transfer contents back to the conical tube. Mix and aliquot 1ml to 1.5 ml microcentrifuge tubes and store at −80° until use.
5. M2 mouse embryo culture medium, available commercially.
6. 5 ml of light mineral oil.
7. Hyaluronidase reconstituted to 10 mg/ml. One vial (Sigma) contains 30 mg; therefore, reconstitute in 3 ml sterile PBS. Peel the metal ring from the vial and remove the rubber stopper. Add 3 ml of PBS to the hyaluronidase and swirl to dissolve. Aliquot 100 μl per microcentrifuge tube and store at −20° until use.
8. Two insulin syringes (1 cc) with a 28 gauge attached needle.
9. Surgical instruments for tissue dissection: one pair three-inch straight scissors, one pair microscissors, four pair watchmakers' forceps #5. Instruments are tip-sterilized in a bead sterilizer prior to use.

10. Four 35 mm plastic petri dishes (do not use tissue-culture treated plastic dishes!).
11. Stereomicroscope with transmitted light and magnification to 80×.
12. Glass embryo transfer capillary and pipetting apparatus.

2.2 CRISPR/Cas9

1. 2 nmol of crRNA for your gene of interest, generated via chemical synthesis from a commercial source, such as Integrated DNA Technologies (IDT).
2. 5 nmol of universal transactivating crRNA (tracrRNA), generated via chemical synthesis.
3. 2 ml microinjection buffer, which is a modified Tris-EDTA (TE) with 0.1 mM EDTA. IDTE can be purchased from IDT.
4. One 0.2 ml PCR tube with cap.
5. Thermocycler.
6. Cas9 protein (New England Biolabs).
7. 1 ml Opti-MEM, room temperature.

2.3 Electroporation

1. Square-wave electroporator, such as the NEPA21 Gene Transfection System.
2. Platinum plate electrode on a glass slide with 5 mm gap.
3. Wash dishes of M2.

2.4 Embryo Culture and Blastocyst Collection

1. KSOM mouse culture medium.
2. Mineral oil or equivalent.
3. 35 mm petri dish.
4. 0.2 ml strip PCR tube.
5. Sterile reverse osmosis (RO) water.
6. −20° freezer.

2.5 Analysis of Activity

1. Blastocyst lysis buffer base (10 ml): 2 ml 1 M Tris–HCl, pH 8, 2 ml 1 M KCl, 40 μl 2% gelatin, 900 μl 10% Tween 20, 5 ml water. Filter through 0.22um syringe filter, then store in sterile 50 ml conical tube. This solution can be stored at room temperature for 6 months.
2. Blastocyst lysis buffer (100 μl): 100 μl blastocyst buffer base, 12.5 μl 10 mg/ml Proteinase K. Prepare this solution immediately prior to use.
3. Thermocycler.

3 Methods

3.1 Embryo Collection

1. Begin superovulation of prepubertal females 3 days before the collection of embryos for electroporation [8]. Slowly thaw an aliquot of PMSG by rolling between your hands. After thawing, load the solution into an insulin syringe and administer 0.1 ml by intraperitoneal injection into each donor female. Shots are administered between noon and 2:00 pm. Return females to their cage.
2. Between 46 and 48 h later, administer 0.1 ml hCG intraperitoneally (i.p.) after thawing in a manner similar to PMSG. After injection, place each female with an individually housed stud male for mating overnight.
3. Females are checked the next morning for a copulation plug (a hardened "plug" of seminal fluid), which indicates the female was mated (*see* **Note 2**). Females are removed from the males' cages.
4. Prepare dishes of culture medium. Add 2 ml M2 medium to one 35 mm petri dish for the harvesting of oviducts. Add 3 ml M2 medium to a second petri dish and to this add a 100 μl aliquot of hyaluronidase (working solution is 0.3 mg/ml) for the collection of embryos. Prepare a wash dish by making three 100 μl drops of M2 medium. Also prepare a holding dish by placing a 100 μl drop of M2 and cover the drop with mineral oil.
5. Sacrifice the donor females according to your Animal Care and Use Protocol as approved by your Institutional Animal Care and Use Committee. Place euthanized females on their backs and spray the abdomen with 70% ethanol (ETOH). Pinch the skin of the abdomen between your thumb and forefinger and lift up; while lifting, make a transverse incision with the straight scissors. Pull the two edges of the incision apart to expose the abdominal wall cleanly. Using the microscissors and forceps, lift and cut the wall transversely to open the abdominal cavity. Move the intestines to one side. The reproductive tract of the female will lie along the back wall of the cavity, just below the kidney. Grasp the uterine horn, lift up, and cut away the connective tissue. Move the forceps to the top of the horn, just below the oviduct and apply light tension. Insert the microscissors in between the oviduct and the ovary and cut. Retain the grasp on the top of the uterine horn and lift, then carefully cut beneath the oviduct. Place the dissected oviduct into the dish of M2 medium (without hyaluronidase). Repeat for the other oviduct in the same female, then repeat the entire process until all 20 oviducts have been harvested.

6. After rinsing the oviducts in M2 (*see* **Note 3**), use a clean pair of forceps to grasp an oviduct and move to the dish of M2 with hyaluronidase. Pin the oviduct to the bottom of the dish with your forceps, then use a second pair of forceps to make a tear in the ampulla and release the cumulus mass (*see* **Note 4**). Remove the empty oviduct and repeat for the remaining oviducts.
7. The hyaluronidase will disperse the cumulus cells and allow the embryos to drop to the bottom of the dish. When the embryos appear smooth and free of cells, pick each embryo up with a pipetting apparatus and a glass embryo transfer capillary and move to the wash dish of M2 drops.
8. Wash the embryos through several drops of M2, then transfer them to the drop of M2 under oil for holding until electroporation (*see* **Note 5**).

3.2 Preparation of CRISPR Reagents

1. CRISPR target sites for the gene and region of interest can be identified by various online bioinformatics resources. This process of selecting proper CRISPR target sites for location and target specificity is beyond the bounds of this chapter [2, 3]. Once the target sites have been identified, CRISPR reagents can be chemically synthesized as crRNAs from a variety of commercial sources, including Integrated DNA Technologies. Alternatively, a guide RNA (gRNA) can be used, either in vitro transcribed or chemically synthesized (*see* **Note 6**).
2. Use the online resuspension calculator on IDT's website to calculate the volume of IDTE required to resuspend the crRNA to 1.0 μg/μl (generally 23.5 μl, assuming 2 nmol of crRNA). Allow the tube to sit at room temperature for at least 1 h. Flick the tube repeatedly to mix and spin briefly to consolidate the volume at the bottom of the tube.
3. Use the resuspension calculator to calculate the volume required to resuspend the tracrRNA to 1.0 μg/μl. Allow the tube to sit at room temperature for at least 1 h, to allow the tracrRNA to become fully suspended. Flick the tube to mix and spin briefly.
4. After both the crRNA and the tracrRNA are resuspended, remove 5 μl of the crRNA and place in the PCR tube. Using a clean pipette tip, remove 10 μl of tracrRNA and add to the 5 μl of crRNA for a total volume of 15 μl.
5. Anneal the crRNA and tracrRNA by placing the tube in the thermocycler and program:
   (a) 95 °C 5 min.
   (b) 92.5 °C 30 s.

   (c) Ramp down temperature 2.5 °C for 30 s per cycle down to 60 °C.
   (d) Ramp down temperature 5 °C for 30 s per cycle down to 25 °C.
   (e) 4 °C until samples are removed.

   Annealed crRNA/tracrRNA may be stored at −80 °C for 6 months.

6. Prepare the Cas9 crRNA/tracrRNA RNP complex. Remove 2 μl of the annealed crRNA/tracrRNA (2 μg of the combined RNAs) and transfer to a 1.5 ml microcentrifuge tube. Add 5.0 μg Cas9 protein directly to this solution (NEB Cas9 protein is at 3.22 mg/ml, so 1.5 μl is required). Pipette gently, quick spin, and allow the guide and Cas9 protein to form ribonucleoprotein complexes for 15 min at room temperature, then dilute the mixture with 96.5 μl Opti-MEM. Mix by pipetting, and the RNP solution is now ready for electroporation and can sit at room temperature for up to 3 h. The final concentration of the combined crRNA/tracrRNA is 20 ng/μl, and Cas9 protein is 50 ng/μl. This final RNP complex is made fresh for each electroporation. Modified prepararation procedures are done if using an in vitro transcribed or synthesized guide RNA (gRNA) (*see* **Note 6**), or if using an ssODN for targeting a mutation via homology directed repair (*see* **Note 7**).

3.3 Electroporation

1. Program the square-wave electroporator. For the NEPA21, the following parameters work well for mouse zygote electroporation: Poring pulse set parameters: 225.0 voltage, 1.0 ms pulse length, 50.0 ms pulse interval, 4 pulses, 10% decay rate, +polarity. Transfer pulse set parameters: 20.0 voltage, 50.0 ms pulse length, 50.0 ms pulse interval, 5 pulses, 40% decay rate, ±polarity.
2. Fill the gap between the plates of the electrode with the entire volume (100 μl) of the CRISPR RNP suspension.
3. Pick up 40 presumptive zygotes with a mouth pipette. Carefully dispense the zygotes so that they are in a single line parallel to the electrode plates (Fig. 1a).
4. Connect the leads and press Start. Bubbles will appear in the well, indicating a current has been generated.
5. Disconnect the leads and pick up the electroporated zygotes via mouth pipette. Return the zygotes to a numbered drop of M2 media.
6. Remove the RNP solution from the electrode well and wash the well three times with Opti-MEM. Fill well with new solution for the next CRISPR reagent to be tested and repeat the process.
7. Electroporate a control CRISPR RNP reagent in the same manner.

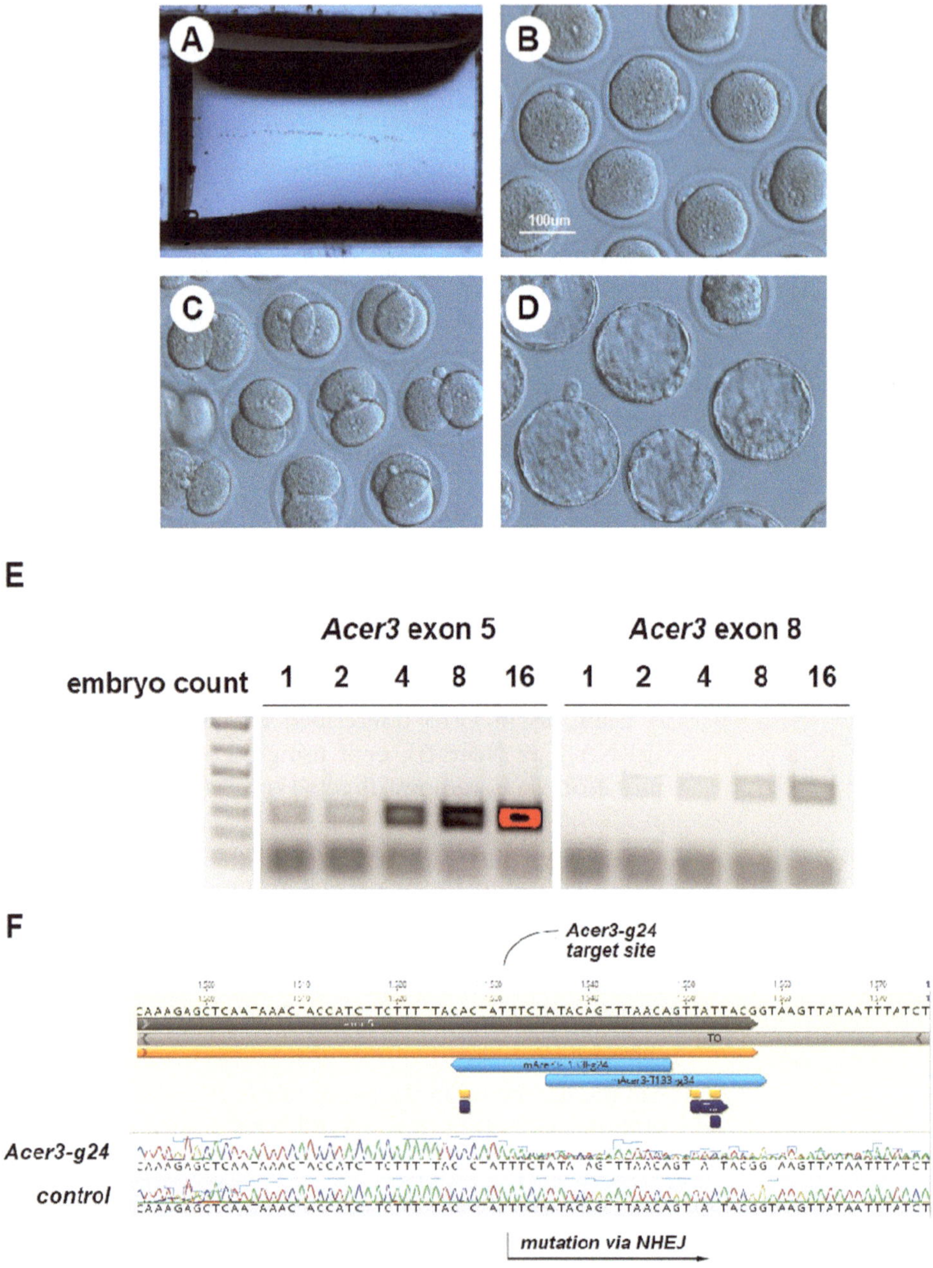

Fig. 1 Summary of CRISPR reagent validation via electroporation of mouse embryos. (**a**) Embryos lined up in electroporator. (**b**) Zygotes immediately after electroporation. (**c**) Culture of the embryos overnight showing successful cleavage to the two-cell stage. Most (>80%) of the electroporated embryos should have advanced to the two-cell stage. (**d**) Culture of the embryos for 4 days showing most embryos (>75%) have advanced to the expanded blastocyst stage. (**e**) Agarose gel of PCR products amplified from genomic DNA from electroporated embryos. For most amplicons, sufficient PCR product can be produced for subsequent procedures. (**f**) Direct sequencing of the PCR product. The presence of mixed electropherograms about the CRISPR target site is indicative of the presence of small deletions and insertions generated via nonhomologous end joining (NHEJ)

3.4 Embryo Culture and Blastocyst Collection

1. Number the bottom of a 35 mm petri dish. Make six 50 μl drops of KSOM and overlay with mineral oil or equivalent. Equilibrate the dish for at least an hour at 5% CO_2, 5% O_2 at 37 °C.
2. Move batches of electroporated embryos (Fig. 1b) through a KSOM wash and into a numbered drop, keeping a record of the CRISPR reagents and drop number and return to the incubator.
3. Score electroporated embryos the next day for development to the two-cell stage (Fig. 1c). If fewer than 80% develop, it may be an early indication of a problem with your electroporation.
4. Culture embryos for a total of 4 days to the blastocyst stage. At least 75% of the embryos should be fully developed to the expanded blastocyst stage (Fig. 1d). Lower percentage of embryos that develop to this extent suggest either an issue with the electroporation or culture conditions, or that the induced mutation potentially impacts preimplantation development.
5. On day five (4 days of culture), aliquot 10 μl of sterile molecular biology-grade water to the bottom of a 0.2 ml PCR strip tube, one tube for each CRISPR reagent tested.
6. Pick up those embryos that have developed to the blastocyst stage via mouth pipette and transfer in a small volume to the 10 μl of water in the PCR tube. It may be possible to visualize the blastocysts in the tube immediately after transfer. Repeat for each group of electroporated embryos into separate PCR tubes.
7. Freeze the PCR tubes at −20 °C until analysis.

3.5 PCR and Analysis of Activity

1. A PCR to amplify the region about the target site needs to be optimized. Details for such optimization are beyond the bounds of this chapter. In brief, a relatively short PCR (300–600 bp) with the CRISPR target site near the center of the PCR product will suffice. This PCR needs to be of sufficient quality to perform direct sequencing, with minimal to no nonspecific bands in the overall product.
2. Thaw collected blastocysts if needed and then add 10 μl of 2× blastocyst lysis buffer to each sample in the PCR tube.
3. Place PCR tubes in the thermocycler.
4. Run the following program to prepare the crude lysate:
   (a) 56 °C for 10 min.
   (b) 95 °C for 10 min.
   (c) 4 °C indefinitely.

5. Flash spin the PCR tubes and then store the crude lysate at −20 °C or use immediately for PCR.
6. Use 3 μl crude blastocyst DNA per PCR.
7. Run the optimized PCR (25 μl reaction) on the crude lysate.
8. Run a small aliquot (10 μl) of the PCR product on an agarose gel to confirm the PCR reaction was successful.
9. Purify the remaining PCR product via column-based purification for commercial Sanger sequencing.
10. Analyze resulting sequence by aligning the electropherogram trace sequence read to both the reference sequence and to the electropherogram trace from control electroporated embryos (Fig. 1f) using a DNA alignment software included in many DNA analysis software such as SnapGene, VectorNTI, or Geneious. Activity of the CRISPR reagent can be confirmed by the presence of mixed PCR products, represented by multiple DNA sequence peaks in the electropherogram near the site of the CRISPR target.

4 Notes

1. F1 embryos, for example from FVBn females mated to C57Bl/6, are convenient for several reasons, including the good response of FVBn females to superovulation over a range of ages and access to male C57Bl/6 breeding stocks. However, polymorphisms between the strains must be considered in the mutation analysis, and F1 embryos will be heterozygous at these polymorphic regions. Small insertion or deletion polymorphisms, particularly in intronic regions, may complicate the use of primary sequencing for the assessment of mutations, due to the presence of mixed electropherograms even in nonmutated mice. If possible, analysis of potential polymorphisms between the strains can be done via analysis of the publicly available genome sequences.
2. Collect embryos from all plugged females. It may also be useful to collect ova from unplugged females and look for the appearance of pronuclei in the cytoplasm, indicating these are likely zygotes. It is possible that the female lost a copulation plug and was actually mated.
3. The less blood you transfer to the embryo collection dish, the easier it is to see the embryos. Washing the oviducts in a dish of M2 prior to collecting the embryos helps to minimize the blood transferred.

4. At day postcoitum (dpc) 0.5 (morning after mating), embryos will be located in the ampulla, which will appear as a swollen and clear section of the oviduct. Embryos are contained within a cumulus mass, which is a mass of cumulus cells that has the appearance of a cloud.
5. Embryos may also be held in KSOM mouse embryo culture medium, but this medium is bicarbonate-buffered and requires CO_2 to maintain the proper pH. KSOM may also be purchased commercially. Prepare a culture dish by placing 50 μl drops of KSOM in a 35 mm plastic petri dish and cover with mineral oil. The culture dish should be allowed to equilibrate at 37° in a humidified chamber with 5% CO_2 (or 5% CO_2 and 5% O_2) for at least an hour before adding embryos.
6. In vitro transcribed or chemically synthesized guide RNA (gRNA) can be used instead of the annealed crRNA:tracrRNA complex. To do so, add 2 μg of gRNA (2 μl of 1.0 μg/μl stock) to 5 μg Cas9 protein (1.5 μl of 3.22 μg/μl stock) as indicated in Subheading 3.2, **step 6**, and incubate for 15 min at room temperature.
7. If a mutation targeted via homology directed repair is desired, a single-strand oligodeoxynucleotide may be included with the RNP mixture by adding 1 μl of a 1.0 μg/μl ssODN suspension directly to the RNP after Subheading 3.2, **step 6**. Dilute with 95.5 μl Opti-MEM as indicated for the final solution for electroporation. The ssODN will have a final concentration of 10 ng/μl.

References

1. Singh P, Schimenti JC, Bolcun-Filas E (2015) A mouse geneticist's practical guide to CRISPR applications. Genetics 199:1–15
2. Labun K, Montague TG, Gagnon JA, Thyme SB, Valen E (2016) CHOPCHOP v2: a web tool for the next generation of CRISPR genome engineering. Nucleic Acids Res 44: W272–W276
3. Concordet JP, Haeussler M (2018) CRISPOR: intuitive guide selection for CRISPR/Cas9 genome editing experiments and screens. Nucleic Acids Res 46:W242–WW45
4. Doench JG, Fusi N, Sullender M, Hegde M, Vaimberg EW, Donovan KF, Smith I, Tothova Z, Wilen C, Orchard R, Virgin HW, Listgarten J, Root DE (2016) Optimized sgRNA design to maximize activity and minimize off-target effects of CRISPR-Cas9. Nat Biotechnol 34:184–191
5. Sentmanat MF, Peters ST, Florian CP, Connelly JP, Pruett-Miller SM (2018) A survey of validation strategies for CRISPR-Cas9 editing. Sci Rep 8:888
6. Kaneko T, Mashimo T (2015) Simple genome editing of rodent intact embryos by electroporation. PLoS One 10:e0142755
7. Anderson KR, Haeussler M, Watanabe C, Janakiraman V, Lund J, Modrusan Z, Stinson J, Bei Q, Buechler A, Yu C, Thamminana SR, Tam L, Sowick MA, Alcantar T, O'Neil N, Li J, Ta L, Lima L, Roose-Girma M, Rairdan X, Durinck S, Warming S (2018) CRISPR off-target analysis in genetically engineered rats and mice. Nat Methods 15:512–514
8. Luo C, Zuniga J, Edison E, Palla S, Dong W, Parker-Thornburg J (2011) Superovulation strategies for 6 commonly used mouse strains. J Am Assoc Lab Anim Sci 50:471–478

Chapter 5

Targeted Mutations in the Mouse via Embryonic Stem Cells

Marina Gertsenstein, Joffrey Mianné, Lydia Teboul, and Lauryl M. J. Nutter

Abstract

Genetic modification of mouse embryonic stem (ES) cells is a powerful technology that enabled the generation of a plethora of mutant mouse lines to study gene function and mammalian biology. Here we describe ES cell culture and transfection techniques used to manipulate the ES cell genome to obtain targeted ES cell clones. We include the standard gene targeting approach as well as the application of the CRISPR/Cas9 system that can improve the efficiency of homologous recombination in ES cells by introducing a double-strand DNA break at the target site.

Key words Mouse, Embryonic stem (ES) cells, Gene targeting, Electroporation, Clustered regularly interspaced short palindromic repeats (CRISPR), CRISPR associated protein (Cas9)

1 Introduction

The isolation and genetic manipulation of embryonic stem (ES) cells is considered to be one of the most important achievements in mammalian biology. Mouse ES cells are characterized by self-renewal, pluripotency, and the ability to contribute to the tissues of chimeric mice, including germ cells [1] such that the ES cell genome can be transmitted to subsequent generations. Gene targeting in mouse ES cells using engineered targeting vectors and the cell-intrinsic homologous recombination repair pathway revolutionized biomedical research and became an essential tool for the generation of genetically modified mice [2]. However, due to the low frequency with which homologous recombination occurs, obtaining and screening for the desired events remains a challenging and time-consuming endeavor.

Introducing a DNA double-strand break (DSB) within the genomic target sequence has been proposed as a strategy to increase the homologous recombination rate at the target site in mammalian cells. This relies on the activity of nucleases directed to the target sequence and has evolved together with the discovery and

Melissa A. Larson (ed.), *Transgenic Mouse: Methods and Protocols*, Methods in Molecular Biology, vol. 2066, https://doi.org/10.1007/978-1-4939-9837-1_5,

development of new genomic engineering tools. While evidence of increased homologous recombination rates has been obtained using I-*Sce*I mega-nuclease [3], zinc finger nucleases (ZFNs) [4], and transcription activator-like effector nucleases (TALENs) [5], the challenge represented by the design and synthesis of such nucleases for each target gene has impeded their common usage. The clustered regularly interspaced short palindromic repeats (CRISPR)/CRISPR associated protein (Cas9) is an RNA-guided nuclease (RGN) that can be targeted to specific genomic sites by simply changing a 20-bp spacer sequence in the single guide RNA (sgRNA) [6]. Cas9 RGN generates DSBs at its target site with high efficiency in the mammalian genome, both in vitro and in vivo [7]. The simplicity of design, synthesis, and delivery of the Cas9 RGN makes it an attractive option for systematic use for enhancing homologous recombination rates in gene targeting experiments [8].

Transfection allows the introduction of foreign DNA into the nucleus of eukaryotic cells. Although new transfection techniques and reagents are constantly emerging, electroporation continues to be widely used to transfect mouse ES cells. Here we describe the standard electroporation protocol as well as the Neon® transfection system that uses an electrified pipette tip as a transfection chamber (http://www.thermofisher.com), the latter in conjunction with Cas9 RGN to introduce DSBs to stimulate homologous recombination.

To produce a genetically modified mouse model, targeted ES cell clones must be able to contribute to the germline of the generated chimeras. ES cells are sensitive to culture conditions and can acquire chromosomal abnormalities during manipulations and prolonged culture that affect this ability. Thus, in addition to meticulous culture conditions to prevent differentiation (described below), chromosome number in targeted clones should be assessed prior to generating chimeras to eliminate aneuploid clones that are unlikely to contribute to the germline [9, 10]. Copy number changes in chromosomes 1, 8, and 11, along with the loss of the Y chromosome from male ES cells, are the most frequently observed aneuploidy events in C57BL/6 ES cells [11]. While karyotyping or chromosome counting of metaphase spreads can identify aneuploidy and has widespread use, we describe alternative protocols for assessing chromosome copy number using real-time polymerase chain reaction (PCR) that is less labor intensive and higher throughput than metaphase spread-based approaches (this article and [11]). If possible, mouse models should be generated from at least two independent targeted clones and backcrossed at least three generations to control for the secondary mutations acquired during ES cell culture [12] and mitigate the risk of making these mutations homozygous during experimental cohort generation.

For decades, ES cell culture relied on medium containing fetal bovine serum (FBS) and a feeder layer of mitotically inactive fibroblasts. It was recently shown that defined serum-free KnockOut™ Serum Replacement (SR)-based medium can improve germline competency of mouse ES cells [13]. Small molecule inhibitors of the Erk and Gsk3 pathways (2i) allowed derivation of mouse ES cells from strains previously considered to be refractory [14]. In particular, C57BL/6 ES cells have been shown to be less stable in culture and less efficient for generating germline chimeras compared to 129 ES cells [15–19]. Several approaches have been used to improve the culture of germline competent C57BL/6 ES cells including medium containing SR [13, 20] or conditioned by ATCC CRL-2647 L-cells [21]. We combined SR-based medium containing 2i signaling pathway inhibitors to culture targeted C57BL/6 ES cells and aggregation method [22] to generate C57BL/6 ES cell chimeras with excellent germline transmission rates [23]. We used the same conditions for the culture of targeted C57BL/6 ES cell clones from international repositories prior to aggregation or microinjection of 8-cell stage outbred albino CD-1 (ICR) host embryos. Based on 742 ES cell clones from nine different parental ES cell lines we achieved 56% clonal germline transmission rate, with 80% from our own C2 ($n = 123$) and 73% from Regeneron VGB6 ($n = 63$) ES cell lines. These ES cell culture conditions (SR + 2i) were adopted by several international centers for high-throughput generation of targeted mouse lines from C57BL/6N-derived ES cells as part of the International Mouse Phenotyping Consortium (www.mousephenotype.org). Protocols for generating chimeras by aggregation or microinjections can be found in the manual by Behringer and colleagues [24].

2 Materials

2.1 Equipment

1. Biosafety cabinet for ES cell manipulations and horizontal laminar flow hood for picking colonies.
2. Aspirator vacuum system with liquid trap.
3. Pipetting aid for 1–25 ml pipettes and pipettors for 2-, 10-, 200-, and 1000-μl tips.
4. Humidified incubator: 37° C, 5–7% CO_2 cleaned regularly.
5. Inverted phase-contrast microscope with 4×, 10×, 20–25× objectives for cell culture observations.
6. Stereomicroscope with transmitted light base for picking colonies.
7. Table-top centrifuge with rotors for 1.5-, 15-, and 50-ml tubes.
8. Freezers: −20 °C, −80 °C, liquid nitrogen storage.

9. Electroporation systems:
   (a) Bio-Rad Gene Pulser® with capacitance extender, 0.4 cm electrode gap cuvettes.
   (b) Neon® transfection system: transfection device, pipette, pipette station, Neon® kit, 100 μl.
10. 8- or 12-channel pipettor with volume adjustable up to 200 μl, sterile pipette tips
11. 8- or 12-channel aspirator system with liquid trap
12. Isopropanol controlled-rate freezing container or styrofoam tube rack with lid for cryovials.
13. Water bath at 37 °C.
14. Hemocytometer or automated cell counter.
15. Incubator set to 56 °C for digestion of ES cells for DNA isolation.
16. Centrifuge with microtiter plate rotor capable of up to 6000 × *g* or vacuum manifold for 96-well plate DNA isolation.
17. Heat block set to 70 °C to preheat DNA elution buffer.
18. Spectrophotometer for quantitation of DNA.
19. Real-time PCR machine and associated software for chromosome copy number assessment.

2.2 Disposable Cell Culture Plastic

1. Sterile tissue culture grade plates: e.g., 6-well, 35-mm, 60-mm, and 100-mm diameter, flat- and U-bottom 96-well plates.
2. Centrifuge tubes: 1.5-, 15-, and 50-ml.
3. Individually wrapped cotton-plugged sterile plastic pipettes (1, 5, 10, and 25 ml).
4. Sterile disposable reagent reservoirs for multichannel pipettor.
5. Cryovials.

2.3 Cells

1. Characterized ES cell lines at early passage can be obtained from the originating investigators or repositories. The parental ES cell line should be expanded with a limited number of passages using culture conditions suggested by the source and tested after expansion for the ability to generate germline chimeras prior to undertaking gene targeting experiments [25].
2. Embryonic stem cells are cultured on mitotically inactivated primary mouse embryonic fibroblasts (MEFs) containing the antibiotic resistance marker to be used for selecting targeted clones in the presence of leukemia inhibitory factor (LIF) or on the mouse fibroblast STO cell line transformed with neomycin resistance (when using neomycin/G418 for selection of targeted clones) and murine LIF genes (SNL). MEFs or SNLs can

be obtained from repositories or commercial sources. MEFs can also be prepared in the laboratory from any strain of transgenic mouse expressing the resistance gene to be used for selecting the altered ES cell genome. A mouse strain Tg (DR4)1Jae/J (stock 003208) resistant to G418 (neomycin), 6-thioguanine, puromycin, and hygromycin is a popular choice as a MEF source. The protocol for isolation of MEFs and their mitotic inactivation is beyond the scope of this chapter, *see* [24] for details. Feeder-free ES cell lines are cultured on gelatin-coated plates.

2.4 Cell Culture

1. The quality of fetal bovine serum (FBS) is one of the most important factors for successful ES cell culture. ES cell-qualified FBS lots are available from suppliers. FBS lots can be tested in the laboratory for plating efficiency, toxicity, and effect on germline potential of ES cells to identify an optimal batch. Once the FBS lot is identified and purchased, it is stored at -80^{0} C.
2. ES-DMEM: Dulbecco's Modified Eagle's Medium (DMEM), 15% FBS, supplements (Table 1) is used for ES cell culture prior to and during electroporation, selection, replica plating, cryo-recovery, and expansion of positive ES cell clones.

Table 1
ES cell culture medium formulations

Reagent	Stock concentration	ES-DMEM		KO-SR + 2i	
		Final [ ]	100 ml	Final [ ]	100 ml
DMEM	–	–	80 ml	–	–
KO-DMEM	–	–	–	–	80 ml
FBS	–	15%	15 ml	–	–
KO-SR	–	–	–	15%	15 ml
GlutaMAX™	200 mM	2 mM	1 ml	2 mM	1 ml
Pen-Strep	10,000 U (μg)/ml	50 U (μg)/ml	0.5 ml	50 U (μg)/ml	0.5 ml
Nonessential amino acids	10 mM	0.1 mM	1 ml	0.1 mM	1 ml
Sodium pyruvate (optional)	100 mM	1 mM	1 ml	1 mM	1 ml
2-Beta-mercaptoethanol	55 mM	0.1 mM	0.2 ml	0.1 mM	0.2 ml
LIF	1×10^{7}U/ml	1000 U/ml	10 μl	500 U/ml	5 μl
Insulin	10 mg/ml	–	–	5 μg/ml	50 μl
PD0325901	5 mM	–	–	1 μM	20 μl
CHIR99021	15 mM	–	–	3 μM	20 μl

3. SR + 2i: DMEM, 15% KnockOut™ Serum Replacement (SR), supplements, 2i (1 μM PD0325901 and 3 μM CHIR9902) (Table 1) is used for ES cell culture prior to generation of chimeras by aggregation or microinjection.
4. MEF-DMEM: DMEM, 10% FBS medium is used for the preparation of MEFs, freezing medium, inactivation of trypsin, and dilution of freezing medium prior to centrifugation. It can be supplemented with 2 mM GlutaMAX and 0.1 mM beta-mercaptoethanol.
5. Freezing medium containing dimethylsulfoxide (DMSO) is prepared fresh and kept on ice prior to use. The concentration of FBS can be increased up to 80–90%.
   (a) 1× (10% DMSO, 10% FBS): 8 ml DMEM, 1 ml FBS, 1 ml DMSO.
   (b) 2× (20% DMSO, 10% FBS): 3 ml DMEM, 1 ml FBS, 1 ml DMSO.
6. MI-DMEM: DMEM, 20 mM HEPES, 15% FBS is used to prepare the single-cell suspension for microinjection.
7. 0.05% Trypsin/EDTA.
8. Accutase.
9. D-PBS, pH 7.0 Ca^{2+}- and Mg^{2+}-free.
10. 0.1% w/v gelatin.
11. Electroporation buffer.
12. The choice of selection reagents (Geneticin (G418), Ganciclovir, FIAU, Hygromycin B, Puromycin) is determined by the targeting construct. The concentration depends on the sensitivity of the particular parental ES cell line and the way the selection is performed (e.g., on MEFs or gelatinized plates). Each lot should be tested in the specific laboratory conditions to determine the lowest dose resulting in 100% death of non-transfected cells within shortest possible time.

2.5 Cas9 RGN Reagent Design and Synthesis

1. Protospacer sequences can be selected using freely available webtools, for example, WTSI Genome Editing (http://www.sanger.ac.uk/htgt/wge/), CRISPOR (http://tefor.net/crispor/crispor.cgi).
2. pSpCas9(BB)-2A-Puro (PX459) v2.0,
3. Targeting construct.

2.6 Targeting Confirmation by Loss-of-Allele Assay

1. Genomic DNA purification kit
2. Taqman or equivalent copy number assay for genomic location disrupted by targeting
3. Taqman or equivalent copy number reference assay.

4. Optical 96-well PCR plates and optical seals.
5. Real-time PCR master mix.

2.7 Chromosome Quality Control by Real-Time PCR

1. Genomic DNA purification kit.
2. Taqman or equivalent assays for genes located on chromosomes 1, 8, 11, and Y.
3. Taqman or equivalent copy number reference assay.
4. Optical 96-well PCR plates and optical seals.
5. Real-time PCR master mix.
6. Control genomic DNA isolated from wild-type mouse from same strain as ES cells using the same protocol as used for ES cells.

3 Methods

3.1 CRISPR Reagent Design and Synthesis

1. Using one of the freely available webtools, scan the target genomic sequence for potential protospacers. Several considerations should be taken into account when selecting protospacer sequences:
   (a) One or more sgRNA(s) can be selected and used during electroporation.
   (b) The sgRNA(s) must be selected in order to allow cleavage of the targeted genomic DNA but not the targeting construct.
   (c) The sgRNA(s) selected should not have potential off-target site within the targeted genome containing less than three mismatches. Ideally at least one, and preferably more, of the mismatches should be within the seed region (11-bp adjacent to the protospacer adjacent motif). (*see* **Note 1**).
   (d) In order to allow efficient transcription of the sgRNA (s) under the U6 promoter in pX459 plasmid, an additional G nucleotide must be added in 5′ of the protospacer sequence(s) if one is not already present.
2. Protospacer sequence(s) can be cloned in pX459 plasmid following Feng Zhang's lab instructions (http://www.genome-engineering.org/crispr/), so both the sgRNA and Cas9 are transcribed from the same electroporated construct.
3. Resuspend pX459 plasmid(s) containing sgRNA(s) in PBS at 1 μg/μl.

3.2 ES Cell Culture

ES cells are very sensitive to culture conditions, it is important to handle them with good aseptic technique and to carefully follow the recommended protocols to maintain their pluripotency and ability to contribute to the germline. ES cells are kept in culture on MEF or gelatin-coated plates at a relatively high density and should be passaged when they reach a subconfluent state of 70–80%, that is, tightly packed intermediate size colonies close to but not touching each other. ES cells should never be seeded too sparsely, that is, when more than 4–5 days are required to reach subconfluency, nor should they ever be grown past 90% confluency as both conditions induce cell differentiation. Growth media is changed every day and cells are dissociated to a single-cell suspension at every passage. The number of passages must be kept as low as possible with a stock of low-passage frozen vials of ES cells stored in liquid nitrogen.

3.2.1 Thawing ES Cells

Fresh medium and MEF-coated or gelatinized plates are prepared in advance with the size corresponding to the number of cells in the cryovial. A vial with 4–5 $\times$ 10^6 cells is typically thawed onto a 60-mm plate. Use face shields, safety gloves, and laboratory-protective apparel when removing vials from cryogenic storage. Loosen the top of the vial to release the pressure from rapidly evaporating liquid nitrogen within the tube to reduce the chance of an explosion. Immediately place the vial on dry ice. The vial should remain on dry ice until ready to thaw. DMSO will begin to damage ES cells soon after thawing, so the steps below should be done quickly and carefully.

1. Remove the vial from dry ice and warm it in a 37 °C water bath for 60–90 s until the ice crystals almost disappear.
2. Dry the vial with a tissue, wipe the outside with 70% ethanol, and place it inside the hood. Using a pipette, aseptically transfer the cell suspension into a 15-ml tube prefilled with 3–4 ml of ES-DMEM to dilute the freezing medium containing DMSO.
3. Pellet the cells by low-speed centrifugation (200–400 $\times$ *g*) for 5 min at room temperature.
4. Carefully aspirate the supernatant leaving a small volume of media behind, flick the pellet to loosen it, and resuspend the cells in fresh ES-DMEM.
5. Plate cell suspension, gently tilt the plate bidirectionally to evenly distribute the cells.
6. Place the plate with cells in the incubator.
7. Change the ES-DMEM the following day, if there are a lot of floating dead cells, rinse with D-PBS before replacing the medium. If the correct procedure was used, the cells should be ready for passage in 2 days.

3.2.2 ES Cell Passage

ES cells should be fed with fresh ES-DMEM every day and ideally ~1–3 h before passage and/or cryopreservation. We routinely use 0.05% trypsin to dissociate cells grown in ES-DMEM and Accutase for cells grown in SR + 2i prior to generating chimeras. Trypsin-treated cells do not attach to gelatinized plates in SR + 2i. Passage with Accutase does not require rinsing with D-PBS, inactivation, or centrifugation, just careful but vigorous pipetting to single-cell suspension and dilution into growth media. ES cells are split at 1:4 to 1:8 ratios depending on their growth rate, ideally every other day. As many as 1–2 × 10^6 cells are typically seeded onto 60-mm plates. Representative 1:6 split ratios are 1 × 35-mm to 1 × 100-mm or 1 × 60-mm to 2 × 100-mm plates.

1. Prepare the necessary number of MEF-coated plates as determined by the split ratio. Mitotically inactivated MEFs are plated at ~3 × 10^4 cells/cm^2 (e.g., ~1.8 × 10^6 cells per 100-mm plate).
2. Aspirate the growth medium from the ES cells and rinse the cells with D-PBS, if needed.
3. Add the appropriate amount of trypsin or Accutase (e.g., 1 ml per 60-mm or 0.5 ml per 35-mm plate), incubate for 3–5 min at 37 °C.
4. Check that the enzymatic treatment has reached the desired level by tapping and swirling the plate to detach the cells from the bottom of the plate.
5. Neutralize trypsin with double volume of ES-DMEM, transfer the cell suspension into a 15-ml tube, and pellet by low-speed centrifugation (200–400 × *g*) for 5 min at room temperature. Aspirate the supernatant, leaving a small volume of media behind, flick the pellet to loosen it, and resuspend the cells in fresh ES-DMEM.
6. If using Accutase, gently pipette the cells while still in Accutase and split cell suspension into new plates containing SR + 2i. If a "string" of cells is formed during enzymatic treatment, it can be left behind; it contains mostly feeders.
7. Count the cells and aliquot the cell suspension onto new plates, for example, ~10 × 10^6 ES cells can be obtained from 60-mm dish, 4–5 × 10^6 cells are typically seeded onto a 100-mm plate.
8. Tilt plates bidirectionally to distribute the cells evenly and place the plates with cells in a humidified incubator at 37 °C, 5–7% CO_2.
9. Change the growth medium (feed the cells) the day after passage.

3.2.3 ES Cell Cryopreservation in Vials

ES cells are frozen in cryovials for long-term storage in liquid nitrogen. The same protocol is used for the wild-type parental ES cell line and targeted clones. Slow cooling at ~1 °C per minute to –80 °C prevents the formation of ice crystals inside the cells and can be achieved by using an isopropanol controlled-rate freezing container or styrofoam tube rack with a lid. It is important to record freezing density as it determines the surface area for thawing. The number of vials to freeze depends on experimental demands; typically a vial containing 3–5 × 10^6 cells is thawed onto a 60-mm plate, for example, 5–6 vials can be frozen from each subconfluent 100-mm plate. When fewer cells are frozen per vial, a smaller surface area is required for thawing, for example, two vials can be frozen from one well of 6-well plate (35-mm diameter) and cells from each vial are thawed on 35-mm diameter surface. All vials should be clearly labeled with alcohol-resistant marker or printed labels indicating the ES cell line, passage number, clone ID, if applicable, date, and cell number or thawing surface.

1. Prepare the necessary volume of 1× freezing medium and keep on ice.
2. Prepare single-cell suspension as for passage described in Subheading 3.2.2.
3. Count the cells; determine the number of vials that can be frozen, for example, 3–5 × 10^6 cells per vial to be thawed onto 60-mm diameter plate.
4. Pellet the cells by low-speed centrifugation (200–400 × *g*) for 5 min at room temperature.
5. Label the appropriate number of vials with all the pertinent information.
6. Remove the supernatant; gently resuspend the cell pellet in the appropriate volume of cold 1× freezing medium to obtain the desired number of cells/ml for each vial.
7. Quickly aliquot 1 ml of the cell suspension into labeled cryovials and place them on ice.
8. Place the vials inside a styrofoam tube rack with lid or an isopropanol controlled-rate freezing container.
9. Immediately place the container in a −80 °C freezer.
10. Transfer cryovials into a liquid nitrogen for long-term storage after 24 h.

3.3 Electroporation of ES Cells

3.3.1 Electroporation with Plasmid Target Vector

Exponentially growing ES cells are used for electroporation and ES-DMEM is changed 2–3 h prior to cell harvest. Targeting vector plasmid DNA is purified using an endotoxin-free kit and linearized with an appropriate restriction enzyme. The linearized vector is then extracted with 25:24:1 phenol/chloroform/isoamyl alcohol, precipitated with ethanol, and dissolved in sterile water at a concentration of 1–2 mg/ml. We use 10–20 μg of linearized targeting vector DNA per ~7 × 10^6 cells in each cuvette.

1. Prepare the necessary number of MEF-coated plates, for example, for electroporation of ~15 × 10^6 cells in two cuvettes we prepare 4 × 100 mm plates at half density (~0.9 × 10^6 MEFs per gelatinized 100-mm plate).
2. Prepare a single-cell suspension from exponentially growing ES cells (e.g., 1 × 100-mm plate) as for passage described in Subheading 3.2.2.
3. Count the cells and pellet them by low-speed centrifugation (200–400 × *g*) for 5 min.
4. Resuspend cell pellet in the appropriate volume of chilled electroporation buffer to have 10 × 10^6 cells/ml.
5. Gently mix 1.4 ml of cell suspension with 20–40 μg of linearized targeting vector DNA without introducing bubbles; aliquot 0.7 ml of the mixture into each of the two cuvettes. Leave on ice for 5 min.
6. Electroporate the cells with the following parameters: 250 V, 500 μF, and ∞ Ohms (Bio-Rad Gene Pulser Xcell), once the pulse is applied, place the cuvette back on ice, record the time constant (typically ~10).
7. After ~5 min on ice, transfer the cells from both cuvettes into a 15-ml tube containing 10 ml of ES-DMEM at room temperature. Gently resuspend the cells and leave them at room temperature for 5 min.
8. Replace the MEF-DMEM in four MEF-coated 100-mm plates with 5 ml of ES-DMEM media.
9. Gently mix the cell suspension and distribute it evenly onto four MEF-coated plates, ~3 ml electroporated cell suspension per plate.
10. Place the plates in the humidified incubator at 37 °C, 5–7% CO_2.
11. The following day, replace ES-DMEM with ES-DMEM plus selection antibiotic as appropriate for the targeting construct. For C57BL/6N-C2 ES cells, we routinely use 100 μg/ml of G418 for ~8–9 days for vectors with neomycin resistance until the colonies are ready to be picked.

3.3.2 Co-Electroporation of Cas9 RGN and a Targeting Construct Using the Neon® Transfection System

ES cells should be expanded as described above to the number required for the experiment, for example, 6×10^5 cells per electroporation. Prepare mitotically arrested MEFs with appropriate antibiotic resistance—one 60-mm plate per construct and two 60-mm plates for controls.

1. Change media on exponentially growing ES cells 2–3 h prior to electroporation to promote growth and cell division.
2. Prepare a single-cell suspension as described in subheading 3.2.2.
3. Plate the cell suspension on a 100-mm plate and incubate for 15–30 min in the humidified incubator. The MEFs in suspension will adhere to the plate in this time while the ES cells will remain suspended.
4. Carefully transfer the supernatant containing the ES cells into a new tube and bring the volume up to 20 ml using ES-DMEM.
5. Count the cells. If the preparation is too concentrated, dilute an aliquot for counting.
6. Pellet the cells by low-speed centrifugation (200–400 × *g*) for 5 min at room temperature.
7. Aspirate the supernatant and resuspend the cells in 10 ml of D-PBS.
8. Transfer the appropriate number of cells into a new tube at a concentration to have 6×10^5 cells per electroporation condition.
9. Pellet the cells by low-speed centrifugation (200–400 × *g*) for 5 min.
10. Prepare and label as many microcentrifuge tubes as electroporations to be performed.
11. Aspirate the supernatant and resuspend the cells in Resuspension Buffer R in order to obtain 6×10^6 cells/ml.
12. Transfer 100 μl of cell suspension (6×10^5 cells) into a labeled microcentrifuge tube and add the appropriate circular DNA construct(s). Control: No DNA added (control for cell death associated with antibiotic selection). Experimental: Add a total of 9 μg of DNA including 3 μg of targeting construct and 6 μg of pX459 plasmid(s) (*see* **Note 2**).
13. Electroporation parameters: pulse voltage set at 1350 V, pulse width set at 30, and number of pulses set at 1 (*see* **Note 3**).
14. Using Neon® 100 μl tips, mix the suspensions containing the cells and DNA by pipetting up and down. Fill up the tip with the suspension avoiding bubbles and place the Neon® pipette in the Neon® pipette station.

15. Perform the electroporation and transfer the cell suspension onto a 60-mm diameter plate containing mitotically arrested antibiotic-resistant MEFs in ES-DMEM without antibiotic. Place the plate in the humidified incubator at 37 °C, 5–7% CO_2.
16. The following day, start antibiotic selection (*see* **Note 4**). The selection is maintained until there are no cells left on control dishes treated with antibiotic and individual clones are ready for picking (typically ~8–9 days after electroporation).

3.4 Picking ES Cell Clones into 96-Well Plates

All sensitive cells should die during selection and the resistant cells (those that have incorporated the targeting construct) will grow into individual colonies visible to the naked eye. Well-separated colonies of similar size and best morphology are selected for picking: smooth with clearly defined edges, compact center, and indistinguishable individual cells. Overgrown large flat colonies with visible cells are differentiating and should be avoided. Colonies can be picked in a biosafety cabinet with a microscope window or a horizontal laminar flow hood using a stereomicroscope with transmitted light or an inverted microscope. The number of colonies to be picked depends on the project. We routinely pick two or three 96-well plates per experiment after standard electroporation or 48–96 clones for Cas9 RGN co-transfection experiments (*see* **Note 5**).

1. The day prior to picking, prepare the appropriate number of MEF-covered 96-well flat-bottom plates.
2. Just before picking colonies, aspirate the MEF-DMEM from 96-well plate with MEFs; replace it with 100 μl/well of ES-DMEM with appropriate selection antibiotic.
3. Aliquot 30 μl of 0.05% trypsin into each well of as many U-bottom 96-well plates as you will be picking colonies for.
4. Aspirate the media from one plate with ES cell colonies leaving ~0.5 ml of ES-DMEM behind; add 8 ml of D-PBS. The number of plates to be handled at a time depends on the number of colonies per plate and the speed of picking. Colonies should not stay in trypsin longer than 30 min.
5. Using a pipettor set to 2 or 10 μl with a sterile tip carefully dislodge the selected colony under the microscope and draw it into the pipette tip with as little volume of D-PBS as possible.
6. Transfer each individual colony into one well of a U-bottom 96-well plate containing trypsin.
7. Using a new tip for each colony, proceed with the rest of the colonies. Keep track of the wells by the position of the corresponding pipette tips in the box.

8. Place 96-well plate with picked colonies to the incubator for 5 min.
9. Working row by row with a multichannel pipettor, add 50–70 μl of selective ES-DMEM to each well containing ES cells.
10. Pipette up and down 4–5 times to disaggregate the cells and transfer the cell suspension to the corresponding row of the prepared 96-well MEF plate containing 100 μl of ES-DMEM with appropriate selection antibiotic per well.
11. Check the cells under the microscope. The wells of U-bottom plate can be rinsed with more ES-DMEM with selection antibiotic that is then added to the corresponding row of 96-well plate (maximum total volume of 200 μl/well).
12. Place the plates in the humidified incubator at 37 °C, 5–7% CO_2.
13. Change the media using a multichannel pipettor the following day and each alternate day (every ~48 h) until the cells are ready for passage. Avoid damaging or lifting cells by pipette tips during media changes.

3.5 Passage of ES Cell Clones in 96-Well Plates

Cells in different wells often do not grow at a synchronous rate. Observe the cells under the microscope and score each well for cell density. The time for the passage should be chosen when about 2/3 of the wells have reached 70–80% confluency. Each 96-well plate can be split into three replica plates: two on MEFs and one on gelatin (*see* **Note 6**). Once two replica plates on MEFs reach 70–80% confluency they are frozen for future recovery and the plate on gelatin is split one more time to get two replica plates for DNA preparation. It is important to maintain antibiotic selection after picking and during replica expansion. However, the concentration has to be decreased for the cells grown on gelatin to avoid cell death.

1. Prepare the required number of flat-bottom 96-well plates: two covered with MEFs and one gelatinized for each 96-well plate ready for passage. To prepare gelatinized plate, 50–100 μl of 0.1% gelatin is added to each well and aspirated after 10–15 min at room temperature.
2. Add 150 μl of ES-DMEM with selection antibiotic per well to new plates and place them back into the incubator.
3. Gently aspirate the media from the 96-well plate to be split, rinse with 150–200 μl/well of D-PBS.
4. Add 50 μl of trypsin per well, incubate at 37 °C for 5 min. The cells should detach with gentle tapping.
5. Add 100 μl of medium to each well to inactivate trypsin. Working row by row with a multichannel pipette, pipette up

and down 4–5 times to produce a single-cell suspension. Transfer 50 μl of the cell suspension into the corresponding wells of each of the three new plates (*see* **Note 7**).

6. Place the plates in the humidified incubator at 37 °C, 5–7% CO_2.
7. Change the media the following day and continue changing every other day until the cells are ready for cryopreservation or passage.

3.6 Freezing ES Cell Clones in 96-Well Plates

Observe the cells in 96-well plate to be frozen under the microscope and score each well for cell density. These records are essential for later recovery of confirmed positive clones.

1. Prepare necessary volume of 2× freezing medium (10 ml per one 96-well plate), keep it on ice.
2. Gently aspirate the medium from the plate, rinse with 150–200 μl/well of D-PBS.
3. Add 50 μl of trypsin per well, incubate at 37 °C for 5 min.
4. Add 50 μl of ES-DMEM, pipette the cells up and down 4–5 times to get single-cell suspension.
5. Working quickly on ice, aliquot 100 μl of 2× freezing media into each well and mix by gentle pipetting.
6. Wrap the plate with one layer of Parafilm and two layers of foil to allow for the gradual slow freezing rate, label the plate.
7. Immediately place it in the −80 °C freezer.
8. Storage at –80 °C longer than 2 months may compromise cell recovery, transfer to liquid nitrogen storage after 24–48 h if possible.

3.7 DNA Isolation and Screening for Correct Targeting

Screening ES cell clones for correctly targeted events can be done in a variety of ways, including long-range PCR and Southern blot. However, these methods are time consuming and can require extensive optimization prior to use. Here we describe methodology using real-time PCR that when coupled with PCR for the target vector (e.g., for the selection cassette), quickly and efficiently identifies cells with correct targeting events. This assay is called the loss-of-allele (LOA) assay for the wild-type allele.

Assay design—Correct targeting will disrupt the target genomic sequence by the addition or removal of wild-type DNA, for example, the removal of a target exon, the insertion of loxP sites, the insertion of a selection cassette. The loss-of-allele assay is designed such that the assay is contained within the deleted segment of the genome or the amplification primers flank the site of insertion and the probe will span the insertion site. Correct targeting will either remove the assay target sequence or disrupt the ability of the probe to bind to the assay sequence. As a result, a correctly targeted ES cell clone will have a copy number of 1 (for autosomal targets) or 0 (for X-linked targets in male (XY) ES cells) compared to wild-type genomic DNA isolated from the cognate background mouse strain.

Quantitative real-time PCR requires purified DNA with A260/280 ratios >1.8 and A260/230 ratios >2. The simplest method to obtain sufficient purified DNA from 96-well plates is to use a kit such as the TissueSpin 96 from Macherey Nagel, though any 96-well kit may suffice. DNA is isolated essentially following kit manufacturer's instructions using either a centrifuge equipped with a microtiter plate rotor or a vacuum manifold from duplicate MEF-free 96-well plates.

Anticipated results—Correctly targeted cell lines will give a copy number of 0.8–1.2 for autosomal targets and 0–0.2 for sex-linked targets. Copy number assessment of the vector, for example, using an assay directed against the selectable marker, will ensure that random integration of the target vector did not occur. Additional quality control to be performed after expansion of targeted clones should include long-range PCR using a vector-specific primer and a gene-specific primer located beyond the end of the target vector homology arm for both 5′ and 3′ sides of the vector. If desired, the long-range PCR products can be sequenced to confirm no rearrangements during targeting and/or additional short-range PCR can be done to confirm allele structure. Droplet digital PCR (ddPCR) can also be used with the advantage that a calibrator wild-type DNA template is not required [11]. However, the equipment required for ddPCR may not yet be widely available and the reagents are somewhat costlier than real-time PCR. Southern blot may be used instead of long-range PCR and/or LOA, however more DNA is required for Southern blotting, optimization of hybridization conditions may be required, and care should be taken to detect and observe extra bands that may indicate random insertion of vector fragments. Finally, short-range PCR for the bacterial vector backbone may be done to ensure no random integration of these sequences.

When Cas9 RGN is used to stimulate homologous recombination, indel mutations may be introduced at sites where nonhomologous end-joining rather than homologous recombination was used to repair the DSB. In some cases, this may obviate the ability to use LOA assays to confirm targeting as indel formation at the target site may also provide LOA assay copy numbers of ~1.0 for autosomal genes and ~0 for sex-linked genes. However, it may be possible to design LOA assays distal from the Cas9 RGN target site, such as when a single gRNA is used to introduce a DSB, but multiple sequence elements (e.g., loxP sites and a neo selection cassette) are part of the desired targeted allele. In cases where LOA assays are not appropriate for identifying targeting events, long-range PCR may be used instead (*see* **Note 8**).

1. Aspirate the ES-DMEM from the 96-well plate and wash cells once with 100 μl D-PBS per well.
2. Add 100 μl of digestion buffer with proteinase K to each well of the duplicate MEF-free DNA plates; place plates in a Tupperware or other sealable container lined with paper towels saturated with water to prevent excess evaporation of the digestion buffer. Seal container and incubate at 56 °C overnight.
3. The next day, using a multichannel pipettor, mix each row of wells by pipetting 2–3 times and then transfer and pool replica wells from the duplicate plates into a 1.2 ml deep-well 96-well plate for further processing.
4. Process per kit instructions and elute DNA with two elution steps using elution buffer preheated to 70 °C and half the recommended elution volume for each elution step, for example, 2 × 50 μl instead of 1 × 100 μl.
5. Quantitate DNA by UV absorbance ensuring that the A260/280 ratio is ≥1.8 and the A260/230 ratio is ≥2.0.
6. Normalize DNA to a working concentration of 5 ng/μl by preparing a replica plate with the appropriate volume of stock DNA from each well and PCR-grade water for 50 μl per well.
7. Thaw the 2× real-time PCR mix on ice. Swirl gently to mix—do not vortex. Add ROX, if necessary (*see* **Note 9**).
8. Completely thaw the TaqMan copy number assay for the target sequence and copy number reference assay (*Tert* or *Tfrc*). Gently vortex the assays to mix and centrifuge briefly to collect contents and clear the lid. If necessary, dilute 60× stock solutions to 20× in PCR-grade water.
9. Using the table below for the single reaction volumes, prepare sufficient PCR mastermix, comprised of all reagents except for the DNA, for all reactions in triplicate—include a no-template control (NTC) and wild-type control (WTC, two copies for autosomal genes, one copy for X-linked genes) (*see* **Note 10**).

2× TaqMan® Genotyping Mix, with ROX	10 μl
20× target gene copy number assay	1.0 μl
20× copy number reference assay (Tfrc or Tert)	1.0 μl
Ultrapure distilled water (DNAse & RNAse free)	3.0 μl
Total volume per reaction	**15.0 μl**

OR

2× KAPA Probe Fast qPCR Mix, with ROX	10.4 μl
20× target gene copy number assay	1.0 μl
20× copy number reference assay (Tfrc or Tert)	1.0 μl
Ultrapure distilled water (DNAse & RNAse free)	2.6 μl
Total volume per reaction	**15.0 μl**

10. Aliquot 5 μl of 5 ng/μl genomic DNA in triplicate for each sample into a 96-well optical PCR plate. Use genomic DNA isolated from a male wild-type mouse of the same strain background as the ES cells as the two- or one-copy control for autosomal or sex-linked genes, respectively (*see* **Note 11**).
11. Dispense 15 μl of the mastermix prepared above into each well/tube of the reaction plate containing DNA using a repeating pipettor and appropriate sterile tip. Take care not to contact the DNA. The tip can be touched to the top of each tube/well if the DNA was carefully dispensed to the bottom of the well/tube. Add mastermix to the nontemplate control (NTC) wells last.
12. Seal reaction plates with optical adhesive seals. Use the seal tool around each well to ensure no reaction components can escape during cycling. Only touch the seal at the edges—residue from gloves can interfere with data collection and invalidate results. If tubes are used, cap firmly with optical caps, and ensure that no gaps are present between the tube and the cap to prevent evaporation of the reagents (*see* **Note 12**).
13. Tap the plate/tubes gently to mix. Centrifuge briefly to collect contents and clear the seal/lid.
14. Set up the Viia7, or equivalent PCR machine, using the program appropriate for real-time PCR analysis for copy number determination. In general, this includes data acquisition for fluorescence of the target gene copy number assay, reference assay (*Tert* or *Tfrc*) and ROX before the first cycle and after each amplification cycle.
15. Place reaction vessel(s) in PCR machine and cycle per machine and assay instructions.
16. Once cycling is complete, use the appropriate software to analyze the data and determine copy number.

3.8 Chromosome Copy Number Assessment

Once correctly targeted clones have been identified, assessment of chromosome copy number will identify aneuploid clones. Metaphase spread chromosome counting and/or karyotyping are described elsewhere [11]. Here, we describe using real-time PCR to determine the copy number of chromosomes 1, 8, 11, and

Y. These quantitative PCRs are set up the same way as those for target confirmation by the LOA assay. As with LOA, ddPCR may be used, if preferred.

Anticipated results—Euploid ES cell lines (male) will have a copy number of 1.8–2.2 for Chr1, 8, and 11, and a copy number of 0.8–1.2 for ChrY. In the event that no targeted ES cell lines have these results, we have obtained germline transmission from a subset of ES cell lines with copy numbers of 1.6–1.8 or 2.2–2.4 for Chr1, 8, or 11 and 0.6–0.8 or 1.2–1.4 for ChrY. Secondary screening by metaphase spread chromosome counting or karyotyping to assess relative populations of aneuploidy and euploid cells in such provisional ES cell lines may be desired.

1. Aliquot 5 μl of 5 ng/μl genomic DNA in triplicate for each sample for each chromosome assay into a 96-well optical PCR plate. Use genomic DNA isolated from a male wild-type mouse of the same strain background as the ES cells as the two- or one-copy control for Chr1, 8, and 11 or Y genes, respectively.
2. Dispense 15 μl of prepared mastermix (*see* Subheading 3.7) into each well/tube of the reaction plate containing DNA using a repeating pipettor and appropriate sterile tip.
3. Seal reaction plates with optical adhesive seals. If tubes are used, cap firmly with optical caps.
4. Tap the plate/tubes gently to mix. Centrifuge to collect contents and clear the seal/lid.
5. Set up the Viia7, or equivalent PCR machine, using the program appropriate for real-time PCR analysis for copy number determination. In general, this includes data acquisition for fluorescence of the chromosome copy number assay (e.g., *Tram1* for Chr1, *Efbn2* for Chr8, *Rnf112* for Chr11, and *Uty* for ChrY), reference assay (*Tert* or *Tfrc*), and ROX before the first cycle and after each amplification cycle.
6. Place reaction vessel(s) in PCR machine and cycle per machine and assay instructions.
7. Once cycling is complete, use the appropriate software to analyze the data and determine copy number.

3.9 Thawing and Expansion of ES Cell Clones from 96-Well Plates

Initial recovery of frozen cells from 96-well plate may require passage at different ratios, including the passage on the same surface (1:1) depending on the number of cells frozen and their growth rate. Refer to the records taken at the time of cryopreservation to determine the optimal surface area on which to plate the thawed cells. Prepare the necessary number of 24-well plates covered with MEFs depending on the number of positive clones.

1. Aspirate the media from MEF-covered wells, replace it with 0.5 ml/well of ES-DMEM, and place the dish back in the

incubator. We thaw the cells in drug-free medium but switch to selection medium as soon the cells recover and start growing.

2. Remove the 96-well plate from the freezer, unwrap it, and place it in the humidified incubator.
3. When the ice crystals have almost disappeared, wipe the outside of the plate with 70% ethanol and place it in the biosafety cabinet.
4. Add 100 μl of warm ES-DMEM to all wells identified for expansion and transfer entire contents of the thawed wells to the labeled wells of 24-well plate containing 0.5 ml of ES-DMEM each. Rinse the original wells with more media and add the rinse to the corresponding wells in 24-well plate.
5. Place the plate with thawed cells into the incubator.
6. Change the growth media after overnight culture and daily thereafter. Use selection medium once the cells recover and start growing.
7. Depending on cell density, the cells may be ready for 1:2–1:4 passage after 2–3 days of culture. If thawed cells formed only a few colonies in the 3–4 days after thawing, they should be trypsinized and plated back into the original well or fresh MEF-coated plates of the same surface area.
8. Typically, at least two vials from one subconfluent 6-well plate should be frozen for each clone; additional cells are pelleted for DNA preparation for quality control and genotype confirmation prior to generation of chimeras.

3.10 Preparation of ES Cells for Generation of Chimeras

It is important to maintain ES cells in optimal culture conditions at all times, but particularly for the generation of chimeras. Ideally, the cells should be at the lowest possible passage and growing exponentially. Targeted C57BL/6 ES cell clones are thawed in advance to ensure at least two passages in SR + 2i media. Prepare the necessary number of MEF-coated plates depending on the number of clones and planned passages, aim to have one subconfluent 35-mm dish per clone (e.g., one well of a 6-well plate) 1 day before planned aggregation or microinjection experiment. Feed the cells with fresh medium 1–3 h prior to their harvest for aggregation or microinjection. The details of aggregation or microinjection techniques can be found in [24].

3.10.1 Preparation of ES Cells for Aggregation

Sparser than usual (e.g.,1:10–1:30) passage on gelatinized plate 1 day before aggregation will produce small colonies of 5–10 cells required for aggregation. This protocol describes the preparation of C57BL/6N-derived ES cells for aggregation. In our experience, 129- and 129 × B6F1 hybrid ES cells do not respond well to culture in 2i-containing media. For clones derived in these strain backgrounds, ES-DMEM should be used instead of SR + 2i medium.

1. Gelatinize two or three 35-mm plates per clone by flooding plates with 0.1% gelatin and removing it after 10–15 min of incubation at room temperature.
2. Remove the medium from growing cells, rinse the cells with D-PBS, add Accutase (e.g., 0.5 ml per 35-mm well), and incubate at 37 °C, 5% CO_2 for 3–5 min.
3. Resuspend the cells in Accutase by gentle pipetting to ensure a single-cell suspension.
4. Plate the cell suspension (e.g., 30 μl, 70 μl, and 100 μl from 0.5 ml) into each of the three 35-mm gelatinized dishes containing 2 ml of SR + 2i media each to obtain different dilutions. Check the cell density under a microscope; add more cells if necessary. The remaining cell suspension can be frozen and/or passaged. SR + 2i media can be added to the original well to serve as an additional backup dilution for aggregation.
5. On the day of aggregation, when zona-free embryos are ready, ES cell colonies are lifted.
6. Remove the medium and rinse the cells with D-PBS. Add the minimum amount of Accutase to cover the cells (e.g., 0.3 ml per 35-mm dish). Leave at room temperature for 1–2 min.
7. Check cells under the microscope, gently swirl the plates to detach the colonies, and tap the dish until cells are lifted and clumps begin floating.
8. Quickly add ES-DMEM to each dish. Do not pipette. If too many cell clumps appear to contain more than 10 cells, gentle pipetting can reduce clump size but take care not to break clumps into a single-cell suspension.
9. Cell clumps are now ready for aggregation and can be picked up directly from the dish. If it is necessary to transport the cells to another laboratory, carefully transfer the clumps into 5 ml tubes without breaking them into single cells. Do not place cell clumps into the incubator; keep them at room temperature during aggregation.

3.10.2 Preparation of ES Cells for Microinjection

One 35-mm plate of subconfluent ES cells plated on gelatin 1 day before microinjection provides a sufficient amount of cells, but two dishes plated at different densities may be convenient to choose the optimum cell concentration. The majority of feeders do not attach in SR + 2i medium when the passage is done on gelatin, as a result the cell suspension prepared for microinjections is not contaminated with feeders.

1. About 45 min before the cells are required for microinjection, remove the medium and rinse the cells with D-PBS. Add Accutase to cover cells (0.5 ml per 35 mm surface), incubate for 3–5 min at 37 °C. Resuspend into single cells by pipetting while still in Accutase. Add FBS containing medium.
2. Centrifuge cell suspension (200–400 × *g*, 5 min), remove the supernatant, and resuspend cell pellet in 1 ml of cold DMEM, 20 mM HEPES, and 15% FBS.
3. Transfer cell suspension to a 1.5-ml microcentrifuge tube and incubate for 20–30 min on ice.
4. Discard the floating dead cells by carefully removing the top 1/4 of the medium (~250 μl) and leaving large clumps and remaining MEFs settled at the bottom of the tube.
5. Transfer the middle 0.5 ml of single-cell suspension to new microfuge tube and place it on ice. These cells are now ready for microinjection.

4 Notes

1. If the cells are used to generate live animals, particular attention should be paid to selecting guide(s) with a low number of potential off-target sites located on the targeted chromosome.
2. Concentration of the different plasmids can be adjusted depending on the cell line and constructs used.
3. Each ES cell line has its own specific set of parameters, see manufacturer's instructions for optimization protocols. We have successfully used the stated parameters for JM8.F6 ES cells.
4. The optimal concentration of antibiotic is dependent on the ES cell line used. We recommend using the selection associated with the targeting construct rather than the one associated with pX459 plasmid in order to select targeting events.
5. Due to the high efficiency at which homologous recombination occurs when co-electroporating targeting construct with pX459, a smaller number of clones might be sufficient to obtain the desired event.
6. With sufficiently robust parental ES cell lines, clones can be split over three multiwell plates on the day of picking.
7. It is possible to reuse the original 96-well plate, in this case only one new MEF and one gelatinized plate are needed and 100 μl of ES-DMEM is added to the remaining 50 μl/well of cell suspension in the original plate.

8. Clones obtained through co-electroporation of Cas9 RGN reagents with a targeting construct are likely to contain small insertions or deletions (indels) at target sites repaired by non-homologous end-joining DNA repair rather than homologous recombination. This should be taken into account when designing the different screening strategies involving loss-of-allele assay.
9. For the analysis software on the Viia7 and other ABI real-time PCR machines, the passive reference dye ROX is used to correct for pipetting errors. The presence of ROX does not negate the need for careful and accurate pipetting. The KAPA Probe-Fast real-time PCR mix may not contain ROX. If needed, add 0.4 μl of the ROX stock solution supplied with the KAPA PCR mix for each 10 μl of KAPA 2× ProbeFast mix. Use 10.4 μl of this PCR mix per 20 μl real-time PCR.
10. Mastermix is prepared immediately before use (never stored) with ~2% additional volume to account for reagent loss during mastermix dispensing. These reagents are light-sensitive, so protect from prolonged exposure to light.
11. Take care to dispense the DNA to the bottom of the tube/well avoiding contact with the sides of the tube/well. Care at this step reduces the chance of cross-contaminating wells during mastermix addition. The tip used to dispense each sample need not be changed between aliquots, provided the tip is prewetted with the genomic DNA prior to drawing up and dispensing the first aliquot to ensure equal amounts of DNA are aliquoted into each tube/well. Alternatively, mastermix can be dispensed first to all tubes/wells and then DNA added directly to the mastermix and mixed by pipetting. In this case, a new tip must be used for each DNA aliquot.
12. Reagent evaporation during PCR can contaminate the PCR machine with nonspecific fluorescence and necessitate expensive and time-consuming cleaning to remove such contamination.

References

1. Bradley A, Evans M, Kaufman MH, Robertson E (1984) Formation of germ-line chimaeras from embryo-derived teratocarcinoma cell lines. Nature 309(5965):255–256
2. Capecchi MR (1989) Altering the genome by homologous recombination. Science 244 (4910):1288–1292
3. Smih F, Rouet P, Romanienko PJ, Jasin M (1995) Double-strand breaks at the target locus stimulate gene targeting in embryonic stem cells. Nucleic Acids Res 23 (24):5012–5019
4. Bibikova M, Carroll D, Segal DJ, Trautman JK, Smith J, Kim YG et al (2001) Stimulation of homologous recombination through targeted cleavage by chimeric nucleases. Mol Cell Biol 21(1):289–297
5. Hockemeyer D, Wang H, Kiani S, Lai CS, Gao Q, Cassady JP et al (2011) Genetic engineering of human pluripotent cells using

TALE nucleases. Nat Biotechnol 29 (8):731–734
6. Jinek M, Chylinski K, Fonfara I, Hauer M, Doudna JA, Charpentier E (2012) A programmable dual-RNA-guided DNA endonuclease in adaptive bacterial immunity. Science 337 (6096):816–821
7. Wang H, Yang H, Shivalila CS, Dawlaty MM, Cheng AW, Zhang F et al (2013) One-step generation of mice carrying mutations in multiple genes by CRISPR/Cas-mediated genome engineering. Cell 153(4):910–918
8. Byrne SM, Ortiz L, Mali P, Aach J, Church GM (2015) Multi-kilobase homozygous targeted gene replacement in human induced pluripotent stem cells. Nucleic Acids Res 43(3):e21
9. Longo L, Bygrave A, Grosveld FG, Pandolfi PP (1997) The chromosome make-up of mouse embryonic stem cells is predictive of somatic and germ cell chimaerism. Transgenic Res 6 (5):321–328
10. Liu X, Wu H, Loring J, Hormuzdi S, Disteche CM, Bornstein P et al (1997) Trisomy eight in ES cells is a common potential problem in gene targeting and interferes with germ line transmission. Dev Dyn 209(1):85–91
11. Codner GF, Lindner L, Caulder A, Wattenhofer-Donze M, Radage A, Mertz A et al (2016) Aneuploidy screening of embryonic stem cell clones by metaphase karyotyping and droplet digital polymerase chain reaction. BMC Cell Biol 17(1):30
12. Liang Q, Conte N, Skarnes WC, Bradley A (2008) Extensive genomic copy number variation in embryonic stem cells. Proc Natl Acad Sci U S A 105(45):17453–17456
13. Bryja V, Bonilla S, Cajanek L, Parish CL, Schwartz CM, Luo Y et al (2006) An efficient method for the derivation of mouse embryonic stem cells. Stem Cells 24(4):844–849
14. Ying QL, Wray J, Nichols J, Batlle-Morera L, Doble B, Woodgett J et al (2008) The ground state of embryonic stem cell self-renewal. Nature 453(7194):519–523
15. Auerbach W, Dunmore JH, Fairchild-Huntress V, Fang Q, Auerbach AB, Huszar D et al (2000) Establishment and chimera analysis of 129/SvEv- and C57BL/6-derived mouse embryonic stem cell lines. Biotechniques 29 (5):1024–8, 30, 32
16. Hansen GM, Markesich DC, Burnett MB, Zhu Q, Dionne KM, Richter LJ et al (2008) Large-scale gene trapping in C57BL/6N mouse embryonic stem cells. Genome Res 18 (10):1670–1679
17. Seong E, Saunders TL, Stewart CL, Burmeister M (2004) To knockout in 129 or in C57BL/6: that is the question. Trends Genet 20 (2):59–62
18. Ward CM, Barrow KM, Stern PL (2004) Significant variations in differentiation properties between independent mouse ES cell lines cultured under defined conditions. Exp Cell Res 293(2):229–238
19. Ware CB, Siverts LA, Nelson AM, Morton JF, Ladiges WC (2003) Utility of a C57BL/6 ES line versus 129 ES lines for targeted mutations in mice. Transgenic Res 12(6):743–746
20. Wong ES, Ban KH, Mutalif R, Jenkins NA, Copeland NG, Stewart CL (2010) A simple procedure for the efficient derivation of mouse ES cells. Methods Enzymol 476:265–283
21. Poueymirou WT, Auerbach W, Frendewey D, Hickey JF, Escaravage JM, Esau L et al (2007) F0 generation mice fully derived from gene-targeted embryonic stem cells allowing immediate phenotypic analyses. Nat Biotechnol 25 (1):91–99
22. Nagy A, Gocza E, Diaz EM, Prideaux VR, Ivanyi E, Markkula M et al (1990) Embryonic stem cells alone are able to support fetal development in the mouse. Development 110 (3):815–821
23. Gertsenstein M, Nutter LM, Reid T, Pereira M, Stanford WL, Rossant J et al (2010) Efficient generation of germ line transmitting chimeras from C57BL/6N ES cells by aggregation with outbred host embryos. PLoS One 5(6):e11260
24. Behringer R, Gertsenstein M, Nagy K, Nagy A (2014) Manipulating the mouse embryo: a laboratory manual, 4th edn. Cold Spring Harbor Laboratory Press, Cold Spring Harbor, NY
25. Kondoh G, Yamamoto Y, Yoshida K, Suzuki Y, Osuka S, Nakano Y et al (1999) Easy assessment of ES cell clone potency for chimeric development and germ-line competency by an optimized aggregation method. J Biochem Biophys Methods 39(3):137–142

Chapter 6

Blastocyst Microinjection with Embryonic Stem Cells

Melissa A. Larson

Abstract

The ability to delete the function of an endogenous gene in the mouse was made possible by the development of embryonic stem (ES) cells, pluripotent cells that retain the ability to develop into all tissues of a developing embryo. The ability to genetically modify these cells followed, allowing targeted mutation of ES cells in vitro and the deletion of specific gene function. However, regardless of the simplicity or complexity of the genetic modification, all ES cells require injection into host embryos to establish pregnancies and result in chimeric mice. Blastocysts are commonly used as the host embryos for this purpose, as it is relatively easy to inject cells into the blastocoel cavity of the developing embryo. This chapter describes the procedure for injection of ES cells into blastocyst stage embryos for the generation of knockout mice.

Key words Embryonic stem cells, Blastocyst, Microinjection, Knockout, Chimera

1 Introduction

It was demonstrated in 1980 that mice could be generated carrying a foreign piece of DNA [1]. This foreign DNA inserts randomly into the genome, and mice carrying such an insertion are referred to as transgenic. The ability to create a targeted mutation in mice was demonstrated in the late 1980s and was dependent upon the establishment of embryonic stem (ES) cells [2–4]. ES cells could be maintained in culture in a pluripotent state, which permits genetic modification and selection of the homologous recombinants. However, ES cells are themselves insufficient to establish a pregnancy, and they require a carrier or host embryo in order to generate a gene-modified mouse [5, 6]. This combination of ES cells and host blastocyst results in a chimeric mouse, as the mouse will have developed from two distinct cell lineages. This chapter describes the generation of chimeric mice following injection of ES cells into host blastocysts.

Melissa A. Larson (ed.), *Transgenic Mouse: Methods and Protocols*, Methods in Molecular Biology, vol. 2066,
https://doi.org/10.1007/978-1-4939-9837-1_6,

2 Materials

2.1 Embryo Collection

1. Five prepubertal embryo donor females. C57BL/6 females should be just 4 weeks of age when superovulation begins (*see* **Note 1**).
2. Five intact stud C57BL/6 males.
3. Pregnant mare's serum gonadotropin (PMSG), reconstituted to 50 IU/ml.
4. Human chorionic gonadotropin (hCG), reconstituted to 50 IU/ml.
5. M2 medium for embryo handling, available commercially.
6. KSOM culture medium, available commercially.
7. Light mineral oil, or equivalent.
8. 30-gauge blunt-end needle and 1 cc syringes.
9. 35-mm plastic petri dishes.
10. Dissecting tools, stereomicroscope, pipetting apparatus, and incubator.

2.2 Blastocyst Microinjection

1. Mouse ES cells that have been maintained in a pluripotent state by appropriate culture.
2. Embryos cultured to the blastocyst stage.
3. 60-mm plastic petri dish.
4. Inverted microscope with magnification to at least 200×.
5. Cooling stage (optional).
6. Micromanipulators, CellTram Air micrometer to control the holding pipette, SAS Air Syringe to control the injection needle (or equivalent).
7. Glass holding capillary with fire-polished end, available commercially.
8. ES cell injection needle, available commercially.

3 Methods

3.1 Embryo Collection

1. Administer 5 IU PMSG intraperitoneally to the five prepubertal females between noon and 2:00 pm 6 days before the planned day of microinjection.
2. Between 46 and 48 h later, administer i.p. injection of 5 IU hCG to the same females. After injection, pair each female with an individually-housed stud male.

3. On day 2.5 (morning of copulation plug is considered day 0.5 post coitum), sacrifice females according to your IACUC-approved Animal Care and Use protocol.
4. With euthanized females on their backs, spray the abdomen with 70% ETOH and open the skin. Open the abdominal cavity with clean instruments (sterilize tips in a bead sterilizer) and locate the reproductive tract lying just beneath the kidney on the back wall of the cavity. Dissect out the entire tract, including the oviducts and uterine horns, and place in a dish of M2.
5. Under the stereoscope, move one tract to a clean dish of M2 with clean instruments and cut to separate each oviduct from the uterine horns.
6. While securing one horn with forceps, fill a 1 cc syringe (with an attached 28-gauge needle) with M2 medium, insert the needle into the horn, and flush gently. Repeat for the other horn and remove flushed horns from the dish (*see* **Note 2**).
7. Attach the 30-gauge blunt-end needle to a 1 cc syringe and fill with M2. Under higher magnification, locate the infundibulum (the opening nearest the ovary) of one oviduct. Pin the oviduct to the bottom of the dish with your forceps, insert the blunt-end needle into the infundibulum, and flush very gently. A swell of embryos and debris should be visible leaving the other end of the oviduct. Repeat for the other oviduct, and for all remaining tracts (*see* **Note 3**).
8. After flushing is complete, pick up embryos at the morula stage of development. Wash through several drops of M2 and place in a drop of KSOM overlaid with oil. Culture overnight under 6% CO_2, 5% O_2 (*see* **Note 4**).

3.2 Blastocyst Microinjection

1. Prepare a microinjection dish by placing a 100 μl drop of M2 slightly off-center in a shallow 60-mm petri dish. Next place a drop of ES cells in suspension in a curved moon shape around the right side of the drop of M2. Overlay both drops in oil (*see* **Note 5**).
2. Move the injection dish to the inverted microscope stage. If a cooling stage is present, set the temperature to 4 °C. Lower the holding pipette and injection pipette into the drop of M2 and situate both pipettes on the same plane of focus.
3. While looking through the microscope at 40×, raise the holding pipette up and back out of the line of sight. Raise the injection needle slightly. Move the stage so that the drop of ES cells is in focus. Lower the injection needle into the drop.
4. Increase the magnification to 200× (or 400× if more comfortable), and focus on individual cells. While applying the slightest amount of suction, pick up those individual cells that appear

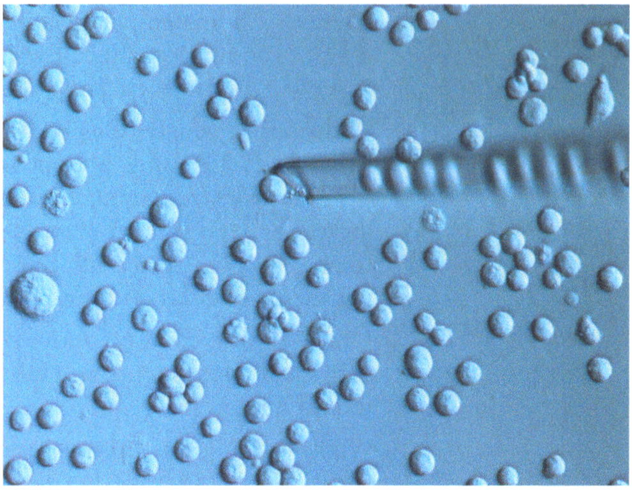

Fig. 1 Embryonic stem cell selection. ES cells are loaded into the injection needle by applying suction. Select those cells with a smooth, round, three-dimensional appearance. This type of cell will almost appear shiny and have a halo of light around the perimeter. Avoid cells that appear flat, rough, or irregular, including large fibroblast cells that may be in the cell preparation. Load one cell after another in a single-file fashion

round and smooth in a single-file fashion. Count as you go, so that you collect as many as 500 ES cells in your injection needle (Fig. 1).

5. Decrease magnification to 40× and move back to the drop of M2. It is possible at this time to introduce the blastocysts into the drop of M2 while the dish is on the inverted microscope. If this procedure sounds too difficult, load the blastocysts into the drop of M2 on the stereoscope prior to placing the injection dish on the inverted stage.
6. Lower the holding pipette into focus and increase magnification to 200×. Secure one blastocyst with the holding pipette by applying gentle suction to the area near the inner cell mass. (It helps to have the inner cell mass (ICM) at either the North or South orientation.) Adjust the focus until you can see junctions between the cells in the trophectoderm. Rotate the blastocyst if needed until a junction is visible (Fig. 2).
7. Test the plane of focus of the injection needle by lightly prodding the visible junction. If the center of the blastocyst gives in slightly, the injection needle is on the correct plane (Fig. 2). If the blastocyst gives in an unbalanced way, then the injection needle is either too high or too low in relation to the cell junction.
8. Once the appropriate plane has been established, give one short, controlled rapid push and punch the needle through

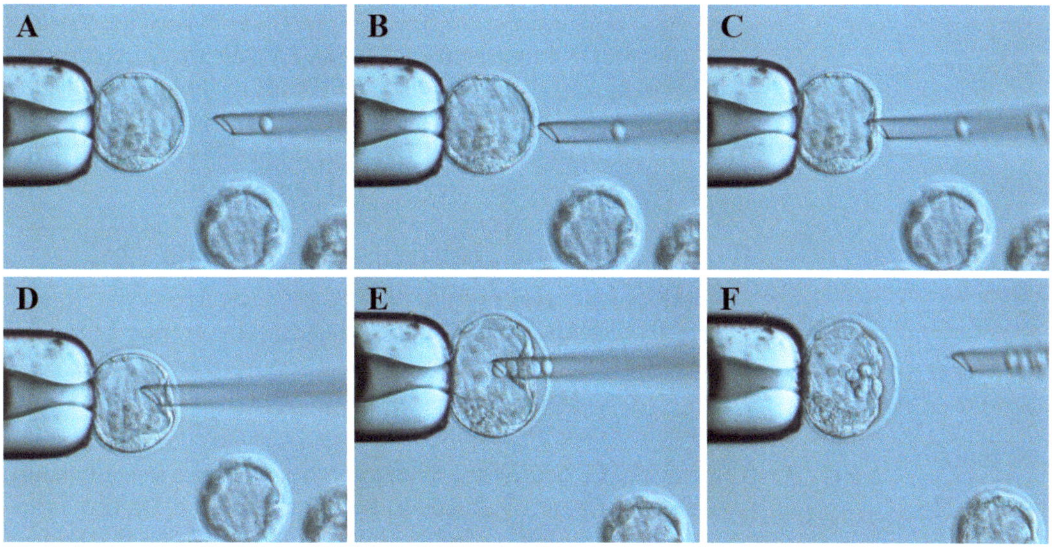

Fig. 2 Injection of embryonic stem cells. (**a**) A blastocyst is secured on a fire-polished holding pipette, and the embryo is situated such that the inner cell mass (ICM) is in the Southern orientation. A gap between the trophectoderm cells has been identified and brought into focus. (**b**) The injection needle is advanced until the tip of the needle lightly prods the zona pellucida surrounding the blastocyst. (**c**) By applying pressure to the injection needle, it is possible to determine if the needle is on the correct plane of focus. In this image, the blastocyst invaginates in the middle, confirming the focal plane. (**d**) The injection needle is pushed further into the blastocoel cavity with a controlled jab. (**e**) Cells are dispensed from the injection needle in single-file fashion into the cavity with controlled pressure, allowing an exact number of ES cells to be deposited. (**f**) After the selected number of cells have been injected, the injection needle is withdrawn, and the blastocyst is allowed to recover with the cells clearly within the blastocoel cavity

the zona pellucida and cell junction, until the bevel of the injection needle is within the blastocoel cavity. Slowly dispense 10–15 cells in the cavity and withdraw the needle (Fig. 2).

9. Move the injected embryo to the top of the drop, pick up another blastocyst, and repeat until the injection needle is empty of cells. If more blastocysts are available, repeat the process of loading the needle with cells and repeating injections.
10. Transfer injected embryos to the oviduct of a 0.5 dpc (days post-coitum) pseudopregnant recipient female or to the uterine horn of a 2.5 dpc pseudopregnant recipient for the generation of chimeric mice.

4 Notes

1. The embryo donor strain chosen depends upon the background strain of the embryonic stem cells. By selecting a host embryo of a different coat color from that of the ES cells, a

mottled or brindle coat color then serves as a visual indicator of the degree of chimerism. In general, ES cells are derived from a 129 background and possess a chinchilla or albino coat; therefore, these ES cells are injected into blastocysts from C57BL/6 females mated to males of the same strain. On the other hand, C57BL/6 ES cells are injected into BALB/C or albino C57BL/6 blastocysts.

2. At 2.5 dpc, the embryos should be at the morula stage of development and present in the oviduct. However, it may vary from female to female, so flushing the uterine horns is a precautionary step to ensure that all embryos are collected. Do not be discouraged if all horns are flushed and no morula are present. Flush the oviducts!
3. Flushing the oviducts is a bit tricky and requires a steady hand. The flushing needle needs to be blunt and small, as the infundibulum is fragile and will readily tear. Once mastered, though, it is quite simple and very effective.
4. It is possible to flush blastocysts from 3.5 dpc donor females on the morning of microinjection. At this point in development, the embryos (blastocysts) should be located in the uterine horns. It should, therefore, only be necessary to flush the horns, although again the location of embryos may differ from female to female (so flushing the oviducts is still advised). However, in our experience, flushing at dpc 3.5 results in fewer embryos collected and many blastocysts without their zona pellucida.
5. As an alternative to creating a separate drop of ES cells adjacent to the drop of M2, it is possible to add ES cells to the drop of M2. The cells will settle to the bottom, and blastocysts may be loaded into the drop on top of the cells. This arrangement minimizes movement between drops, but it does make the injection dish messy.

References

1. Gordon JW, Scangos GA, Plotkin DJ, Barbosa JA, Ruddle FH (1980) Genetic transformation of mouse embryos by microinjection of purified DNA. Proc Natl Acad Sci U S A 77:7380–7384
2. Thompson S, Clarke AR, Pow AM, Hooper ML, Melton DW (1989) Germ line transmission and expression of a corrected HPRT gene produced by gene targeting in embryonic stem cells. Cell 56:313–321
3. Koller BH, Hagemann LJ, Doetschman T, Hagaman JR, Huang S, Williams PJ et al (1989) Germ-line transmission of a planned alteration made in a hypoxanthine phosphoribosyltransferase gene by homologous recombination in embryonic stem cells. Proc Natl Acad Sci U S A 86:8927–8931
4. Zijlstra M, Li E, Sajjadi F, Subramani S, Jaenisch R (1989) Germ-line transmission of a disrupted beta 2-microglobulin gene produced by homologous recombination in embryonic stem cells. Nature 342:435–438
5. Gardner RL (1985) Clonal analysis of early mammalian development. Philos Trans R Soc Lond B Biol Sci 312:163–178
6. Beddington RSP, Robertson EJ (1989) An assessment of the developmental potential of embryonic stem cells in the midgestation mouse embryo. Development 105:733–737

Chapter 7

Generation of Large Fragment Knock-In Mouse Models by Microinjecting into 2-Cell Stage Embryos

Bin Gu, Marina Gertsenstein, and Eszter Posfai

Abstract

Large fragment knock-in mouse models such as reporters and conditional mutant mice are important models for biological research. Here we describe 2-cell (2C)-homologous recombination (HR)-CRISPR, a highly efficient method to generate large fragment knock-in mouse models by CRISPR-based genome engineering. Using this method, knock-in founders can be generated routinely in a time frame of about two months with high germline transmission efficiency. 2C-HR-CRISPR will significantly promote the advancement of basic and translational research using genetic mouse models.

Key words CRISPR-Cas9, 2-Cell stage, Knock-in, Homologous recombination

1 Introduction

CRISPR-Cas9-based genome editing technology has transformed mouse genetics by largely eliminating the need to carry out laborious and time-consuming gene modifications in mouse embryonic stem (ES) cells, and has allowed direct generation of gene-modified mouse models through embryo manipulations [1]. As originally reported, the generation of the insertion or deletion (indel)-based gene knockouts through the non-homologous end join (NHEJ) pathway, and targeted point mutations or small insertions through the short single-stranded oligo (ssODN)-mediated single-strand template repair (SSTR) pathway are generally efficient and can be achieved by microinjection or electroporation at the zygote stage [2–5]. However, although initially reported to be relatively efficient [6], large fragment knock-in by homologous recombination after microinjecting circular plasmid donors has proven to be very variable in efficiency. As large fragment knock-in is required for the generation of many types of important genetically modified alleles,

Dr. Posfai as Co-Corresponding Author

Melissa A. Larson (ed.), *Transgenic Mouse: Methods and Protocols*, Methods in Molecular Biology, vol. 2066,
https://doi.org/10.1007/978-1-4939-9837-1_7,

such as reporters, conditional alleles, exon replacements, and humanized alleles, more efficient and robust technologies are required. Given this importance, many technologies based on different mechanistic rationales including EasiCRISPR [7], Tild-CRISPR [8], and PITCh [9] have been developed and shown to achieve large fragment knock-in in different contexts, but each with specific limitations.

In this chapter, we describe 2-cell (2C)-homologous recombination (HR)-CRISPR [10]. By microinjecting CRISPR-Cas9 reagents into 2-cell stage mouse embryos, we take advantage of the uniquely long G2 phase of the cell cycle (10–12 h) at this stage, and harness the stronger activity of the homologous recombination pathway during this cell cycle phase to achieve large fragment knock-in with high efficiency. We were able to consistently achieve 10–50% efficiency by microinjecting Cas9 mRNA/sgRNA and circular plasmid donors with long homology arms (~1 kb arm length) in a locus-dependent manner. Furthermore, by implementing Cas9-monomeric streptavidin (mSA) mRNA/sgRNA and using biotinylated PCR products with long homology arms as donors [10, 11], the knock-in efficiency could be further improved, in some cases close to 100%. 2C-HR-CRISPR has been successfully implemented to generate reporter alleles, conditional alleles, large transgene insertions into safe harbour sites, and exon/whole gene swapping alleles and therefore has been demonstrated to be an important advancement in mouse genome editing.

In this chapter, we will cover technical details on designing and preparing reagents for 2-cell microinjection and the process of 2-cell microinjection.

2 Materials and Equipment

1. Plasmids: pCS2+ Cas9, pCS2+Cas9-mSA, pX330-U6-Chimeric_BB-CBh-hSpCas9, easyFusion plasmids, knock-in donor plasmids.
2. Cloning kit and reagents: InFusion HD Cloning Kit, CloneAmp HiFi PCR Premix, NucleoBond® Xtra Maxi Kit, NucleoSpin® Gel and PCR Cleanup.
3. In vitro transcription and cleanup of RNA: mMESSAGE, mMACHINE® SP6 Transcription Kit, MEGAshortscript T7 Kit, RNeasy Mini Kit.
4. Microinjection buffer: 10 mM Tris, pH 7.4–7.5, 0.15 mM EDTA. Prepared in double distilled water from 1 M Tris–HCl stock and 0.5 M EDTA stock. For example, for 10 ml microinjection buffer, use 100 μl 1 M Tris–HCl pH 7.5 and 3 μl 0.5 M EDTA. Use microinjection buffer to dilute Cas9/Cas9mSA mRNA, guide RNA, and donor plasmid for injection and filter through Spin-X centrifuge tube-filters 0.45 μm.

5. Anti-vibration isolation table or Air Table (e.g., Sutter Instrument https://www.sutter.com/).
6. Inverted microscope (e.g., Leica DMIRB) or similar with 4×, 20×, and 40× objectives, Differential Interference Contrast (DIC Nomarski) or Hoffman Modulation Contrast (HMC) optics.
7. Micromanipulators (e.g., Leica mechanical manipulators) with a dual instrument holder on the holding side. For micromanipulators without the dual mounting option (e.g., Eppendorf TransferMan®), a ground wire electrode from the amplifier can be attached directly to the microscope stage.
8. Holding system (e.g., Eppendorf CellTram® 4r Air, Oil or similar).
9. Injection system (e.g., Eppendorf FemtoJet® 4i or similar).
10. Intracellular amplifier, e.g., World Precision Instruments (WPI) (https://www.wpiinc.com/) Cyto721 (discontinued) or Duo773 with the following accessories: 3259 footswitch, 13,620–2 mm pin to socket cable, two electrode holder handles 5444, MEH7W10 electrode holder for holding side, MEH2SFW10 electrode holder for injection side, Luer Adapter ¼ 28 female-to-female luer to connect with FemtoJet tubing, e.g., https://www.idex-hs.com/store/luer-adapter-1-4-28-female-to-female-luer-tefzelr-etfe.html. WPI is developing a new model MICROePORE™ dedicated specifically for microinjection of embryos.
11. Microinjection chamber. DIC optics requires use of glass, either standard slide, coverslip, or glass depression slide (e.g., VWR 470235–728). Other options include Nunc™Lab-Tek™ chamber slide with removable wells (e.g., ThermoFisher Scientific 124453PK-1 well) or plastic dish with glass bottom (e.g., WillCo-Wells® GWSB-5040; https://willcowells.com/). Plastics can be used with HMC optics.
12. Pipette puller Flaming/Brown™ style (e.g., Sutter Instrument P-87 https://www.sutter.com/).
13. Microforge (e.g., Narishige; http://narishige-group.com/).
14. Borosilicate glass capillaries with Filament (for injection pipettes): e.g., Sutter BF100-78-1, without Filament (for holding pipettes) B100-75-10 or similar. Alternatively, pipettes can be purchased (e.g. Biomedical Instruments; https://biomedical-instruments.com/).
15. Microloader™, Eppendorf 930001007.
16. Micropipettor.
17. Stereomicroscope with transmitted light for embryo manipulations (e.g., Leica MS5, MS6, or similar).
18. Humidified incubator at 6% CO_2, 37°C.

19. Petri dishes of various sizes for embryo collection and culture.
20. Syringes 1 cc, 30 G 1/2 needles. The sharp tip is polished on the sharpening stone or sand paper to make a needle suitable for flushing the oviducts.
21. Bunsen or alcohol burner for pulling embryo-manipulating pipettes.
22. Pasteur pipettes or glass microcapillaries (e.g., Drummond 1-000-0400 or 1-000-0500) drawn by hand over the flame and broken flat with ID~100 μM and connected to an aspirator mouthpiece (e.g., Drummond 2-000-000 or Sigma A5177) using elastic rubber tubing.
23. Instruments (e.g., Fine Scientific Tools (https://www.finescience.ca) or similar): scissors, fine forceps.
24. Embryo media (e.g., http://www.cytospring.com or http://www.emdmillipore.com): KSOM with amino acids ($KSOM^{AA}$) and M2.
25. Embryo tested light mineral oil (e.g., Millipore ES-005C).
26. Hormones for superovulation: Pregnant mare's Serum Gonadotropin (PMSG) (e.g., Prospec HOR-272), human Chorionic Gonadotropin (hCG) (e.g., Millipore 23074). Hormones are kept at −80°C in aliquots containing 50 IU. The aliquot is diluted in 1 ml of cold PBS immediately prior to injections.

3 Methods

3.1 Targeting Construct Design (Fig. 1) and Cloning (Plasmid and Biotinylated Linear dsDNA)

1. Select 5′ and 3′ homology arms around targeting site. Recommended arm length: ~1 kb each. We observed a decrease in targeting efficiency if targeting arms were shorter than 1 kb, but saw no significant increase when arm length was >1 kb. Select donor vector to clone in arms; *see* available donor vectors with different reporter tags (Addgene EasyFusion Vectors from Rossant Lab: https://www.addgene.org/Janet_Rossant/).
2. Clone in 5′ and 3′ arms from selected genomic sites: Order primers to amplify arms from mouse genomic DNA or bacterial artificial chromosomes (BACs) with appropriate extensions (as described in the InFusion HD Cloning manual) and insert arms into donor vectors using InFusion cloning. Sequence-verify targeting constructs.

3.2 Preparation of Donor Template for Microinjection (Plasmid and Biotinylated Linear dsDNA)

1. Preparation of circular plasmid donor: Produce donor plasmid in DH5alpha bacteria and isolate plasmid with an endotoxin free maxiprep kit; for example, NucleoBond® Xtra Maxi Kit. Maxi prep ends with precipitation of plasmid. Precipitate again, as follows, to further purify the plasmid for microinjection. Add 1/tenth volume 3 M NaOAc and mix. Add × 2 volume 100%

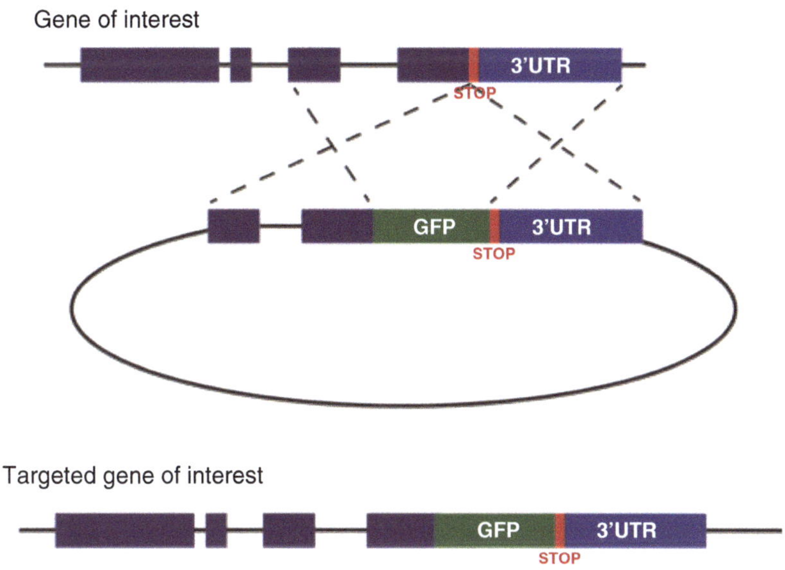

Fig. 1 Example of targeting arm design for C-terminal fusion reporter

EtOH, mix, and keep at −20 °C for a few hours or overnight. Spin and carefully discard supernatant. Wash two times with × 2 volume of 75% EtOH, spin, and discard supernatant. Air dry and resuspend in nuclease-free H_2O, ideally at ~1 μg/μl concentration. Store at −20 °C.

2. Preparation of biotinylated PCR donor: If using EasyFusion donor vectors, use universal biotinylated PCR primers (ordered from Sigma) to amplify any targeting template. Forward oligo: 5′Bio-CTCGAGGTCGACGGTATCGATAA; Reverse oligo: 5′Bio-ATAGGGCGAATTGGAGCTCCAC. Alternatively, design and order your sequence-specific 5′ biotinylated PCR primers. PCR-amplify donor fragment using CloneAmp PCR mix. PCR reaction for each donor: 4 μl template plasmid (10 ng/μl), 4 μl Bio Forward Primer, 4 μl Bio Reverse Primer, 38 μl ddH_2O, 50 μl 2× CloneAmp. PCR programme: 98°C 30 s, 98°C 10 s, 60°C 10 s, 72°C Xs (depends on Amplicon size, generally 5–10 s/kb for CloneAmp), cycle 2–4 35 cycles, 72°C 2 min, 4°C hold. Run 2 μl on a 1% agarose gel to make sure it is a single band and the correct size. Cleanup PCR product using the PCR cleanup protocol in the PCR and Gel purification Kit (e.g., Macherey-Nagel 740609). Precipitate PCR product: Add 1/tenth volume 3 M NaOAc and mix. Add 2× volume 100% EtOH, mix, and keep at −20 °C for a few hours or overnight. Spin and carefully discard supernatant. Wash with 2× volume of 75% EtOH, spin, and discard supernatant. Air dry and resuspend in nuclease-free H_2O, ideally at ~200 ng/μl concentration. Store at −20 °C.

3.3 Guide RNA (sgRNA) Design

We used the Zhang lab's CRISPR design tool (http://crispr.mit.edu). But recently, this server is closed down, and therefore we now switched to CRISPOR (http://crispor.tefor.net/crispor.py) for guide RNA designing. Copy and paste ~100 nt genomic sequence surrounding the desired insertion site into CRISPR design tool, choose mouse, and submit to retrieve sgRNA options.

Things to consider when choosing a guide: Try to choose a guide where the insertion site is within the PAM or the seed sequence. This way the guide will only cleave the genomic site but not the repair template. If this is not possible, you have to mutate one of the Gs in the PAM sequence of the targeting template to introduce a silent mutation (still coding the same amino acid). Cas9 is predicted to cleave 3–4 bps upstream of the PAM sequence. We try to choose a guide with a PAM as close as possible to the desired insertion site; however, we have successfully used guides with the PAM 29 nts away from the insertion site.

3.4 Making Single Guide RNA (sgRNA)

1. Cloning of sgRNA: Order sense and antisense oligos coding the sgRNA sequence from Sigma, anneal, and clone into pX330 vector between BbsI restriction sites as previously described [12].
2. Designing primers for In Vitro Transcription (IVT) of sgRNA: We use PCR products as templates for the IVT. The forward primer introduces the T7 promoter into the sgRNA DNA template and has a sgRNA-specific sequence (N_{18-20}). The reverse primer is universal for all sgRNAs. Design and order the following oligos from Sigma: T7-sgRNA F: 5-′-*TAATACGACTCACTATAGG*N_{18-20}-3′, T7-sgRNA R: 5-′-AAAAGCACCGACTCGGTGCC-3′. The minimum promoter sequence needed for efficient transcription is underlined. N_{18-20} represents your guide-specific target sequence. Note that the last two Gs in the T7 promoter are the first bases that are incorporated into the RNA.
3. Production of a T7-PCR template by PCR for IVT: 10 ng pX330-sgRNA, 1 μl Phusion HF Taq polymerase, 10 μl 5× buffer, 1 μl 10 mM dNTPs, 1 μl T7-sgRNA F oligo (10 μM), 1 μl T7-sgRNA R oligo (10 μM), ddH_2O to 50 μl. PCR program: 98 °C 30 s, 98 °C 10 s, 58 °C 30 s, 72 °C 30 s, ×35 cycles, 72 °C 2 min, 4 °C hold. Run 5 × 50 μl PCR reactions and pool them at the end. Check 5 μl PCR reaction on gel to ensure you have a single band. Purify rest of the pooled PCR product using PCR cleanup protocol of NucleoSpin® Gel and PCR Cleanup (e.g., Macherey-Nagel 740609). Avoid gel purification because it decreases IVT efficiency. Elute with 30 μl RNase-free water. Quantify using Nanodrop.
4. In vitro transcription (IVT) of sgRNA: The sgRNA template is in vitro transcribed using the Ambion MEGAshortscriptT7 Kit

(AM1354), according to the manual. For optimal results, we use 1 μg PCR template and incubate at least 4 h or overnight. IVT reaction: 2 μl Polymerase enzyme mix, 2 μl each ATP, CTP, GTP, UTP, 10 μl 10× buffer, 1 μg T7-PCR template, add ddH_2O to 20 μl. Incubate >4 h at 37 °C. After the IVT reaction, add 1 μl Turbo DNase, gently mix, and incubate 15 min at 37 °C to remove the DNA templates.

5. sgRNA purification: For RNA cleanup use RNeasy Mini Kit (Qiagen 74106). Elute in 30 μl RNase-free water. Quantify using Nanodrop, yield should be between 700 and 3000 ng/μ l. Aliquot sgRNA and store at −80 °C.

3.5 Making Cas9 or Cas9-mSA mRNA for Microinjection (see Note 1)

1. In Vitro Transcription (IVT) of Cas9 or Cas9-mSA mRNA: Cas9 or Cas9-mSA has been cloned into the pCS2+ IVT plasmid, which contains an SP6 promoter. Linearize plasmid using NotI-HF. Run 5 μl on gel to check complete linearization. Purify rest by ethanol precipitation (as described before). Resuspend in RNase-free water. Quantify using Nanodrop. Cas9 is then in vitro transcribed using the Ambion mMessage mMachine SP6 Kit according to the manual. We typically run a double reaction with 2 μg linearized template and incubate at least 4 h or overnight. IVT reaction: 20 μl NTP/CAP, 4 μl 10× buffer, 2 μg linearized Cas9-pCS2, ddH_2O to 40 μl. Incubate >4 h at 37 °C. After the IVT reaction, add 2 μl Turbo DNase, gently mix, and incubate 15 min at 37 °C.
2. Cas9 mRNA purification: For RNA cleanup, use RNeasy Mini Kit (Qiagen 74106). Elute in 30–40 μl RNase-free water. Quantify using Nanodrop, and yield should be between 600 and 1200 ng/μl. Run 2 μl on gel to check integrity. Aliquot Cas9 mRNA and store at −80 °C.

3.6 Preparing Reagents for Microinjection

Prepare injection mix fresh before microinjection. When using Cas9 mRNA: 100 ng/μl Cas9 mRNA, 50 ng/μl sgRNA, 30 ng/μl circular plasmid donor. When using Cas9-mSA mRNA: 75 ng/μl Cas9-mSA mRNA, 50 ng/μl sgRNA, 20 ng/μl BioPCR donor. We typically make 15–25 μl microinjection mix. Filter mix using Spin-X centrifuge tube-filters 0.22 μm (e.g., Costar 8160), keep on ice until use.

3.7 Superovulation and Embryo Collection

1. We use CD-1 (ICR) (Charles River) outbred albino stock as embryo donors as well as pseudopregnant recipients. Donor females at 5–6 weeks of age are injected with 5 IU of PMSG at noon or 12:30 pm followed 47 h later by 5 IU of hCG and mated right away with proven breeder males. Room light cycle: 5 am/on, 5 pm/off. In the next morning, the females are checked for the presence of a vaginal copulation plug as evidence of successful mating.

2. Donor females are euthanized by cervical dislocation at 1.5 post-coitum (dpc) as early as possible in the morning aiming to complete microinjections by noon. Collected 2-cell stage embryos are kept in microdrops of $KSOM^{AA}$ covered by embryo-tested mineral oil at 37 °C in 6% CO_2 with groups of ~30 embryos withdrawn for microinjection as needed. The details of animal colony, embryo collection, and culture are beyond the scope of this chapter. Refer to Behringer et al. for details [13].

3.8 Microinjection into Two-Cell Stage Embryos

Electrophysiological systems utilizing negative capacitance have been routinely used for microinjection of a variety of reagents into mammalian oocytes as well as pre-implantation and post-implantation embryos in developmental biology studies [14–19]. These systems allow the needle to pierce the cell membrane with minimal physical trauma based on the negative capacitance pulse applied to the membrane. We used this system (currently discontinued WPI Cyto 721 model) for the microinjection into the cytoplasm of 2-cell stage mouse embryos and describe its application here [10]. The Micro-ePore intra-cellular amplifier system has been developed by WPI to assist 2-cell cytoplasmic injection and is available now (https://www.wpiinc.com/micro-epore-pinpoint-cell-penetrator). It is also possible to inject the reagents into the nuclei of 2-cell stage embryos using conventional methods routinely used by transgenic facilities, without the assistance of amplifier. The concentrations may need to be adjusted in this case.

1. Preparation of the micropipettes: We use a programmable Sutter Puller (model P-87) equipped with 2-mm square box filament. With this instrument and BF100-78-1 glass capillaries, the ramp value is 565; it may change with the new lot of capillaries or filament change. We use the following program setting: Heat 555, Pull 70, Velocity 70, Delay 120, and aim to have the pipette with an opening and short taper (~6 mm). It is steeper than long and thin micropipettes often used for pronuclear microinjection. Injection pipettes are pulled immediately prior to use. Holding pipettes can be purchased or pulled from glass capillaries without the filament and cut flat using the microforge at ~50–100 μm. The tip of the pipette is then fire-polished to 15–20 μm. The pipette can be bent at ~15° if desired. The holding pipette can be reused until it clogs or breaks.
2. Setting up the microinjection system with amplifier (Fig. 2): Place a drop of M2 media (~0.6 ml) on the glass slide located on the microscope stage. If a smaller drop (~100 μl) is used, it has to be covered with embryo-tested oil to prevent evaporation. The holding pipette and the ground/reference electrode need to be mounted on the holding side, in the dual

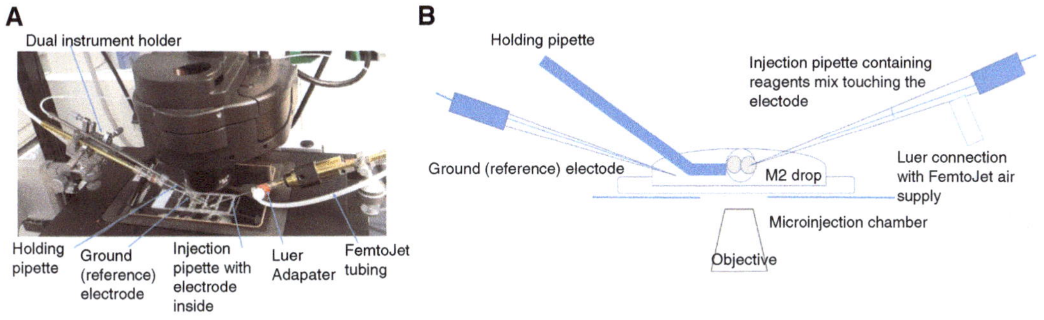

Fig. 2 A picture (**a**) and a diagram (**b**) of the Microinjection Setup: A glass slide serves as a microinjection chamber and contains a large drop (~100 μl) of M2 media that can be covered with embryo-tested oil. The electrode inside the microinjection pipette containing the reagents mix is connected with the amplifier. The electrode holder of the microinjection pipette is connected with the air pressure supply, e.g., FemtoJet via Luer Adapter and tubing. The holding pipette is connected with an oil or air system syringe that allows applying suction pressure to the embryo. The ground reference electrode connected to the amplifier is located inside the medium drop containing the embryos during the microinjection session

instrument holder. Place the holding pipette into the drop until it is just touching the bottom surface. Place a glass capillary with manually broken tip into the holder over the ground/reference electrode and lower it into the M2 drop. The electrode has to touch the medium containing the embryos but its location does not need to be in the operator's view. If the micromanipulator is not equipped with a dual instrument holder, the ground/reference electrode can be attached directly to the microscope stage. Backload the microinjection pipette using Eppendorf Microloader tips attached to the automatic pipettor with 4–5 μl of reagent mix under the stereomicroscope. Make sure there are no air bubbles in the capillary; flick gently if necessary. Place the microinjection pipette into the injection-side single-instrument holder over the electrode and tightly secure it. Make sure the solution touches the wire. Lower the pipette into the drop. The pipette holder outlet is equipped with Luer Adapter connecting it with the FemtoJet injection tubing. Make sure this connection is tight. Switch on the amplifier from "standby" to "operate" mode and proceed with microinjection.

3. Microinjection procedure: Transfer a group of two-cell stage embryos (about 30 embryos) into the M2 drop on the microscope stage. Pick up one embryo at a time by applying suction to the holding pipette system; position it horizontally (Fig. 3). Bring the tip of the microinjection pipette near the zona pellucida. To penetrate the membrane, press the "tickler" pedal of the amplifier and keep it pressed until the injection pipette tip has penetrated the membrane—the pipette tip should "slide in" smoothly (*see* **Note 2**). Once in the cytoplasm, release the "tickler" pedal and apply injection pressure from the

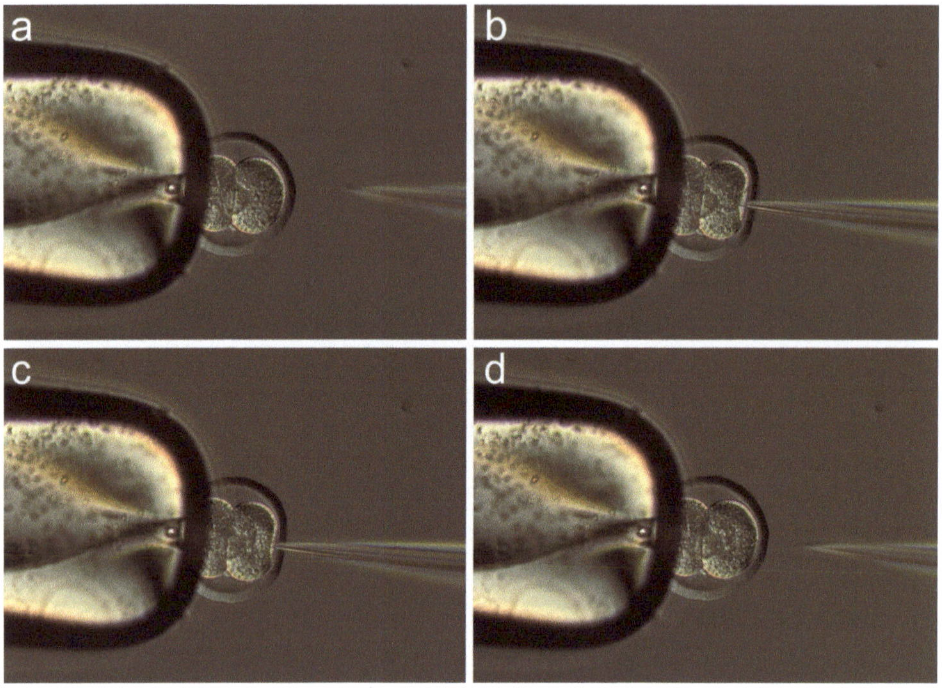

Fig. 3 Example of microinjection into the cytoplasm of a blastomere at the 2-cell stage. (**a**) The 2-cell embryo is horizontally oriented with the holding and injection pipettes. (**b**) Insertion of the injection pipette using negative capacitance. (**c**) Injection of reagents into the cytoplasm—successful injection is visible by a "burst" in the cytoplasm. (**d**) Retrieval of the injection pipette. Following injection of one blastomere, the embryo is flipped and the other blastomere is injected in the same manner

FemtoJet using a computer mouse or foot pedal. You should see a clear burst in the cytoplasm that indicates successful injection. Withdraw the injection pipette from the embryo. Release the embryo, rotate it to bring the second blastomere next to the microinjection pipette tip, and inject the same way. When all embryos in the group are injected in both blastomeres, remove them from the M2 drop, rinse through a few drops of CO_2-equilibrated $KSOM^{AA}$, and return to the incubator. Proceed with the rest of the embryos until all are injected (*see* **Note 3**).

3.9 Embryo Transfers

Microinjected embryos can be transferred into the oviducts of 0.5 dpc pseudopregnant females immediately after microinjection or cultured overnight in microdrops of $KSOM^{AA}$ covered with embryo-tested mineral oil at 37 °C in 6% CO_2 and transferred the next morning at the morulae stage (*see* **Note 4**). The details of surgical embryo transfer are described in Behringer et al. [13].

3.10 Screening and Quality Control for Positive Founders

Founders are screened with long-range genotyping PCR with one primer within the insert and one primer outside the homology arm as previously described [1]. Positive founders are further validated by Sanger sequencing of the knock-in allele and copy number analysis [20].

4 Notes

1. Cas9 RNP: We were successful in generating knock-in alleles by micro-injecting Cas9 RNP, assembled by either spCas9 protein (Thermo Fisher) and in vitro transcribed sgRNA or spCas9 protein (Synthego) and modified synthetic sgRNA (Synthego), with circular donor plasmids or biotinylated PCR donors.
2. Difficulty in penetrating the plasma membrane of 2-cell stage embryos when using intra-cellular capacitor: Make sure the electro-circuit is closed. On Cyto721, an alarm will sound if the circuit is not closed (e.g., the wire is not touching the solution inside the capillary).
3. Embryo lysing after injection: (a) Injection volumes are too high. Adjust injection pressure or use injection needles with smaller opening. (b) Impurity of injection reagents. Further purify regents: For RNA, make sure absorbance ratios are the following: A260/280 > 2.0 and A260/230 > 2.2. For DNA, make sure A260/280 is between 1.8 and 1.9 and A260/230 is ~2.0.
4. Low pup numbers: Further purify the reagents (*see* **Note 4**). Transfer the embryos immediately after microinjections instead of overnight culture.

Acknowledgements

This work was supported by CIHR (FDN-143334) and Genome Canada and Ontario Genomics (OGI-099). The authors wish to thank Dr. Janet Rossant for her guidance and support during the development of 2C-HR-CRISPR and critical discussion and comments.

References

1. Yang H, Wang H, Jaenisch R (2014) Generating genetically modified mice using CRISPR/Cas-mediated genome engineering. Nat Protoc 9:1956–1968. https://doi.org/10.1038/nprot.2014.134
2. Wang H et al (2013) One-step generation of mice carrying mutations in multiple genes by CRISPR/Cas-mediated genome engineering. Cell 153:910–918. https://doi.org/10.1016/j.cell.2013.04.025
3. Yang H et al (2013) One-step generation of mice carrying reporter and conditional alleles by CRISPR/Cas-mediated genome engineering. Cell 154:1370–1379. https://doi.org/10.1016/j.cell.2013.08.022
4. Modzelewski AJ et al (2018) Efficient mouse genome engineering by CRISPR-EZ technology. Nat Protoc 13:1253–1274. https://doi.org/10.1038/nprot.2018.012
5. Qin W et al (2015) Efficient CRISPR/Cas9-mediated genome editing in mice by zygote electroporation of nuclease. Genetics 200:423–430. https://doi.org/10.1534/genetics.115.176594
6. Cohen J (2016) 'Any idiot can do it.' Genome editor CRISPR could put mutant mice in

everyone's reach. Science. https://doi.org/10.1126/science.aal0334

7. Quadros RM et al (2017) Easi-CRISPR: a robust method for one-step generation of mice carrying conditional and insertion alleles using long ssDNA donors and CRISPR ribonucleoproteins. Genome Biol 18:92. https://doi.org/10.1186/s13059-017-1220-4
8. Yao X et al (2018) Tild-CRISPR allows for efficient and precise gene knockin in mouse and human cells. Dev Cell 45:526–536. e525. https://doi.org/10.1016/j.devcel.2018.04.021
9. Nakade S et al (2014) Microhomology-mediated end-joining-dependent integration of donor DNA in cells and animals using TALENs and CRISPR/Cas9. Nat Commun 5:5560. https://doi.org/10.1038/ncomms6560
10. Gu B, Posfai E, Rossant J (2018) Efficient generation of targeted large insertions by microinjection into two-cell-stage mouse embryos. Nat Biotechnol. https://doi.org/10.1038/nbt.4166
11. Ma M et al (2017) Efficient generation of mice carrying homozygous double-floxp alleles using the Cas9-Avidin/biotin-donor DNA system. Cell Res 27:578–581. https://doi.org/10.1038/cr.2017.29
12. Ran FA et al (2013) Genome engineering using the CRISPR-Cas9 system. Nat Protoc 8:2281–2308. https://doi.org/10.1038/nprot.2013.143
13. Behringer R, Gertsenstein M, Nagy K, Nagy A (2014) Manipulating the mouse embryo: a laboratory manual, 4th edn. Cold Spring Harbor Laboratory Press, Cold Spring Harbor
14. Balakier H, Pedersen RA (1982) Allocation of cells to inner cell mass and trophectoderm lineages in preimplantation mouse embryos. Dev Biol 90:352–362
15. Lawson KA, Pedersen RA (1987) Cell fate, morphogenetic movement and population kinetics of embryonic endoderm at the time of germ layer formation in the mouse. Development 101:627–652
16. Wianny F, Zernicka-Goetz M (2000) Specific interference with gene function by double-stranded RNA in early mouse development. Nat Cell Biol 2:70–75. https://doi.org/10.1038/35000016
17. Chazaud C, Yamanaka Y, Pawson T, Rossant J (2006) Early lineage segregation between epiblast and primitive endoderm in mouse blastocysts through the Grb2-MAPK pathway. Dev Cell 10:615–624. https://doi.org/10.1016/j.devcel.2006.02.020
18. Swann K, Campbell K, Yu Y, Saunders C, Lai FA (2009) Use of luciferase chimaera to monitor PLCzeta expression in mouse eggs. Methods Mol Biol 518:17–29. https://doi.org/10.1007/978-1-59745-202-1_2
19. Posfai E et al (2017) Position- and hippo signaling-dependent plasticity during lineage segregation in the early mouse embryo. Elife 6. https://doi.org/10.7554/eLife.22906
20. Gertsenstein M, Nutter LMJ (2018) Engineering point putant and epitope-tagged alleles in mice using Cas9 RNA-guided nuclease. Curr Protoc Mouse Biol 8:28–53. https://doi.org/10.1002/cpmo.40

Chapter 8

Embryo Transfer Surgery

Melissa A. Larson

Abstract

Embryo transfer surgery is an essential step in the process of generating gene-modified mice, regardless of whether the embryos were modified by DNA, RNA CRISPR components, or embryonic stem cells, or whether they were microinjected or electroporated. Transfer is also a necessary step for assisted reproductive techniques such as rederivation and reanimation by in vitro fertilization. The manipulated embryos must be returned to the reproductive tract of a pseudopregnant recipient female mouse, wherein the transplanted embryos develop to term. Embryos may be transferred to either the oviduct or uterine horn, depending upon the developmental status of the embryos and the stage of the recipient. This chapter will describe the process of transferring embryos surgically to a recipient female.

Key words Embryo transfer, Surgery, Mouse, Recipient, Rederivation, Pseudopregnant, Vasectomized male

1 Introduction

While many studies can be conducted with cells and embryos in culture, the generation of a gene-modified mouse requires the development of manipulated embryos to term and subsequent delivery of live-born pups. Similarly, re-establishment of a mouse line following import or in vitro fertilization requires transfer of embryos to recipient females. While the genotype of the recipient females is not important, it is critical that the recipient females be pseudopregnant. Pseudopregnancy is a phenomenon mimicking pregnancy induced by the release of prolactin following cervical stimulation after mating to sterile (vasectomized) males [1]. The prolactin surge is responsible for establishment of the corpus luteum, which produces progesterone and maintains the pregnancy [2].

The morning following mating to vasectomized males, females exhibiting a copulation plug are considered to be 0.5 days post-coitum (dpc). The dpc stage of the recipient mouse determines the stage of embryos that may be transferred [3]. Embryos at all

Melissa A. Larson (ed.), *Transgenic Mouse: Methods and Protocols*, Methods in Molecular Biology, vol. 2066,
https://doi.org/10.1007/978-1-4939-9837-1_8, © Springer Science+Business Media, LLC, part of Springer Nature 2020

preimplantation stages of development (one-cell to blastocyst) may be transferred to the oviduct of a 0.5 dpc recipient female. Blastocysts may also be transferred to the uterine horn of a 2.5 dpc recipient. This chapter will describe both the surgical transplantation of manipulated embryos to the oviduct and the uterus of a recipient female.

2 Materials

1. Colony of 20 breeding-age females (*see* **Notes 1** and **2**).
2. Colony of 20 vasectomized, individually-housed stud males (*see* **Note 3**).
3. Dedicated surgery space and/or a laminar flow hood.
4. Avertin anesthetic, 100% stock: resuspend 5 g tribromoethanol in 5 ml tertiary amyl alcohol. Keep in the dark at 4°C.
5. Avertin anesthetic, 2.5% working solution: dilute 250 ml 100% Avertin stock in 9.75 ml sterile PBS, pass through a syringe filter, and store in a brown bottle at 4°C (*see* **Note 4**).
6. Analgesic, as required by your IACUC-approved Animal Care and Use Protocol. 0.3 mg/ml Buprenorphine: dilute one ampule (1 ml) to 0.03 mg/ml in 10 ml final volume.
7. Clippers, Povidone swabs, and ETOH swabs.
8. Bead sterilizer.
9. Surgical instruments, including 4 pairs of #5 watchmakers' forceps, 3″ straight scissors, microscissors, bulldog clamp, wound clips, and applicator.
10. Embryos to be transferred in M2 medium (*see* **Note 5**), pipetting apparatus.
11. 28-gauge needle.
12. Water-circulated heating pad.

3 Methods

3.1 Oviductal Transfer

1. The day before recipients are needed, mate available females to vasectomized males overnight.
2. The next morning, check females for copulation plugs. Separate those females possessing a plug into a clean cage. Plugged females are considered 0.5 dpc.
3. Weigh female(s) and record weights. Anesthetize one mouse with 0.015 ml Avertin per gram of body weight administered intraperitoneally and return her to the cage. Wait for the anesthetic to take effect; she will be in the correct plane of anesthesia when she does not flinch following a toe pinch.

4. Remove the female from the cage and place her on a clean surface. Clip the hair on the back of the mouse above the left flank. Scrub with Povidone and clean Povidone with 70% ETOH wipes (three).
5. Administer analgesic pre-surgically. Inject 0.1 cc Buprenorphine subcutaneously at the back of the neck.
6. Move the mouse to the laminar flow hood, adjacent to the microscope with her head facing North. Using tip-sterilized instruments (sterilize tips in bead sterilizer at 250 °F for 20 s and allow the instruments to cool on a sterile surface), pick up a pinch of skin above the left flank at the point of the hip. Make a small dorsal incision parallel to the backbone with the straight scissors. With clean forceps and the microscissors, grasp and cut open the body wall through the incision in the skin. Use a second pair of clean forceps and open a hole in the fascia beneath the body wall. Hold open the side of the body wall and carefully look for a fat pad that should be to the left and down. Remove cautiously with forceps. The correct fat pad will be attached to the ovary, followed by the oviduct and uterine horn. The tissues will appear pink. If you remove coiled tubes that appear brown or green, you have the intestines! Clip the fat pad with the bulldog clamp and lay over the back of the mouse.
7. Load embryos to be transferred into your transfer pipette with your pipetting apparatus. (A mouth pipette is most convenient.) In order to control the embryos, load medium, a small bubble of air, medium, air, medium, air, and then the embryos, respectively, in a small volume of medium. Load a small bubble of air after the embryos (*see* **Note 6**). Hang your pipetting apparatus over the binoculars or set to the side.
8. Move the mouse to the stage of the stereoscope on a Kimwipe and focus on the oviduct at 8×. Note the presence of a swollen ampulla (the distended, clear portion of the oviduct downstream of the infundibulum). If the ampulla is not distended, the female is unlikely to become pregnant, regardless of how successful your transfer may be. A distended ampulla indicates the female ovulated and was in estrus, but it is possible that she was not in estrus and still exhibited a copulation plug. In this case, skip this female and move on to another candidate recipient.
9. Locate the infundibulum (the opening of the oviduct just beneath the ovary) through the bursa (the membrane surrounding the ovary and top of the oviduct). Adjust the magnification if necessary. Grasp a pinch of the bursa above the infundibulum with clean forceps and repeat with a second pair of forceps. Slowly and carefully pull apart the bursa to create a hole, while avoiding blood vessels, if possible.

10. Pick up your pipetting apparatus. With one hand holding the pipette, guide the pipette into the infundibulum and clasp the oviduct gently around the transfer pipette with forceps. Gently blow to dispense the embryos in the oviduct. Hopefully, bubbles will appear in the ampulla, indicating your embryos were properly deposited.
11. Withdraw your pipette. Remove the bulldog clamp and move the recipient to the side of the microscope. Return the reproductive tract to the body cavity, and close the skin with a wound clip.
12. Move female to a clean cage situated on the warming pad and allow to recover.

3.2 Uterine Transfer

1. Three days prior to the planned transfer of blastocysts, mate females to vasectomized males.
2. Check for the presence of copulation plugs the next morning. Separate plugged females to a clean cage. Plugged females are considered 0.5 dpc.
3. On the day of blastocyst transfer, the females are considered 2.5 dpc. Prepare the females in the same manner as for oviductal surgery (**steps 3–6**).
4. Load blastocysts into the transfer pipette as in **step** 7.
5. Move the mouse to the stage of the stereomicroscope and focus on the top of the uterine horn at low magnification. Using the 28-gauge needle, make a hole on the top of the uterine horn, being careful to avoid blood vessels (if possible). Be certain the needle has penetrated through to the lumen of the uterus (the needle should move freely while inserted through the wall). Make note of the location of the hole.
6. Pick up the loaded transfer pipette and guide the tip through the hole in the uterine wall. The pipette should move freely within the lumen. While watching the air bubbles in the pipette, gently blow until the bubbles move down the pipette. This is an indication that the blastocysts have been expelled into the lumen.
7. Return the tract to the body cavity and close the incision.
8. Allow mouse to recover on a warming pad until ambulatory.

4 Notes

1. While the genotype of the recipient female is not important, it is important that she be of a strain with good mothering skills. CD-1 female mice are often used, as they are outbred, inexpensive, and generally bear large litters when mated naturally. The

hope is that this tendency to large litters will make the CD-1 female well suited to care for a litter of valuable pups. However, CD-1 females will also become large if kept too long, and females over 35 g will make the embryo transfer surgery very difficult due to the presence of large fat pads on the ovary.

2. If resources are limited and it is not possible to maintain a colony of 20 outbred females for recipients, it is possible to synchronize just a few females by superovulating. Administer 5 IU Pregnant Mare's Serum Gonadotropin (PMSG) intraperitoneally to five females of breeding age three days before recipients are needed. Forty-six to 48 h later, administer 5 IU human Chorionic Gonadotropin (hCG) by i.p. injection and pair synchronized females with vasectomized males for mating overnight. Check for plugs the next morning. Note that preparing recipient females in this manner, while possible, is not widely practiced. Alternatively, it is possible to determine those females in proestrus by the appearance of the vulva, and only mate those females that are likely to come into estrus overnight. This practice requires some experience and may not be absolute. The most reliable method of estrous cycle stage detection is vaginal cytology; while obviously more time-consuming than setting up random matings, this method is reliable and worth pursuing should only a few mice be available [4].
3. The genotype of the stud vasectomized males is not important, and again, outbred CD-1 males are often used. Males may be vasectomized in-house or simply ordered from the vendor already vasectomized. (Charles River will perform this surgery, for a fee.) A record of the plugging dates should be recorded for each male, and males should be replaced when no plugs are noted for a month or so (males are usually replaced at 1.5 years of age).
4. It is certainly possible to use other anesthetics if required by your IACUC. Avertin is popular and allows the surgeon to move the mouse from station to station for clipping, scrubbing, opening the body wall, and performing the transfer under a microscope; however, it is not a pharmaceutical-grade anesthetic and there have been reports of irritation and lethality if Avertin crystallizes [5]. Therefore, it is important to make the working solution in small batches, to filter and to replace frequently. Even so, Avertin will depress body temperature of the anesthetized mouse, so warming is important for the fairly long recovery time to sternal recumbency. Alternatively, inhalation anesthesia results in almost immediate recovery, although the requirement for nose cones and tubes is cumbersome.

5. M2 medium is a HEPES-buffered medium for handling embryos out of the incubator. Just prior to embryo transfer surgery, remove embryos from the incubator and transfer to a drop of M2 without oil overlay. Wash the transfer pipette through several other drops of M2 to remove oil. M2 is available for purchase commercially.
6. A transfer pipette should be a pulled capillary with a long working area. Even after loading three bubbles for the transfer, medium should not pass the shoulder where the pulled pipette meets the unpulled capillary. Work with medium just in this pulled part of the pipette.

References

1. Gunnet JW, Freeman ME (1984) Hypothalamic regulation of mating-induced prolactin release. Effect of electrical stimulation of the medial preoptic area in conscious female rats. Neuroendo 38:12–16
2. Osada T, Watanabe G, Sakaki Y, Takeuchi T (2001) Puromycin-sensitive aminopeptidase is essential for the maternal recognition of pregnancy in mice. Mol Endocrinol 15(6):882–893
3. McLaren A, Michie D (1956) Studies on the transfer of fertilized mouse eggs to uterine foster mothers 1. Factors affecting the implantation and survival of native and transferred eggs. J Exp Biol 33:394–416
4. Byers SL, Wiles MV, Dunn SL, Taft RA (2012) Mouse estrous cycle identification tool and images. PLoS One 7(4):e35538. PMID: 22514749
5. Zeller W, Meier G, Burki K, Panoussis B (1998) Adverse effects of tribromoethanol as used in the production of transgenic mice. Lab Anim 32 (4):407–413

Chapter 9

Nonsurgical Embryo Transfer Protocol for Use with the NSET™ Device

Barbara J. Stone

Abstract

Genetically modified embryos must be transferred to a suitable female recipient for development to pups. Nonsurgical embryo transfer is a fast and efficient method used to deliver blastocyst stage embryos to the uterine horn of recipient females. The efficiency of recovery of pups after nonsurgical embryo transfer is similar to the efficiency after surgical transfer. However, nonsurgical transfer eliminates the pain and distress caused by the surgical procedure and provides a refinement in accordance with Russel and Burch's "3Rs" (The principles of humane experimental technique. Methuen & Co., London, 1959), an ethical framework for animal research. This method is also useful for rederivation of mouse strains. Rederivation is important for either removal of potential pathogens from an incoming mouse strain after shipping, or within a facility to obtain a clean mouse colony.

Key words Nonsurgical, Embryo, Embryo transfer, NSET, Anesthesia, 3Rs, Rederivation, Transgenic, Animal welfare

1 Introduction

While all animal research carries the responsibility of regard for [1] animal welfare , one of the simplest refinements available is exclusion of a surgical procedure. Strategies for genetic manipulation in mice require the transfer of embryos to a recipient female for development of those embryos to produce genetically modified offspring. In order to eliminate the need for surgical embryo transfer, a nonsurgical embryo transfer device (NSET™) and methods have been developed for its use in mice [2]. This technology eliminates a surgical procedure and the potential post-surgical complications which may occur. The procedure has been shown to be less stressful to mice than a comparable surgical procedure with anesthesia [3]. The NSET device for embryo transfer has been used for many studies of mouse reproduction [4–10] and to transfer genetically altered embryos [2, 11–13].

Melissa A. Larson (ed.), *Transgenic Mouse: Methods and Protocols*, Methods in Molecular Biology, vol. 2066,
https://doi.org/10.1007/978-1-4939-9837-1_9, © Springer Science+Business Media, LLC, part of Springer Nature 2020

The NSET device consists of a tapered catheter which can be inserted past the cervix to deposit embryos into the uterine horn. The cervix must be dilated for passage of the catheter to occur. Conveniently, the cervix is appropriately dilated at 2.5 days post-coitum (2.5 dpc). This is the optimal time for transfer of blastocysts, the embryonic stage found at 3.5 dpc (e3.5), to the mouse uterine horn. The procedure is minimally invasive and does not require the use of anesthesia or analgesia. The embryo transfer procedure can be performed in under 30 s from selection of the recipient to the return of the animal to its cage. No post-procedure monitoring is required. Compared to surgical embryo transfers, this procedure dramatically reduces the time and expertise required for the production of genetically modified or rederived mice.

2 Materials

1. Pseudopregnant female mice to serve as embryo recipients. Females should be at least 8 weeks old and mated to a vasectomized male overnight 3 days prior to the embryo transfer procedure to produce a pseudopregnant recipient at 2.5 dpc. Mating activity must be confirmed by the presence of a vaginal plug.
2. Blastocyst transfer is recommended. Embryos can be obtained directly from a female donor. Embryos can also be transferred after cryopreservation, in vitro fertilization (IVF), pronuclear injection, ES cell injection, or culture.
3. NSET device (ParaTechs Cat # 60010).
4. Speculum (two sizes are included with each NSET device).
5. P-2 pipette.
6. Incubator or slide warmer, 37 °C.
7. Stereomicroscope with transmitted and reflected illumination source with 20× and 40× magnification.
8. 60 mm tissue culture dishes.
9. Embryo handling pipette assembly.
10. Wire-topped cage.
11. Goose-neck lighting (optional).
12. M2 medium equilibrated to 37 °C.

3 Methods

All applicable international, national, and institutional guidelines for the care and use of animals must be followed. Sterile technique should be used. All embryos and media should be kept at 37 °C.

1. Just prior to the embryo transfer procedure, place one 20–40 μl drop of M2 onto the inside surface of a culture dish lid (*see* **Note 1**).
2. Under the microscope while using reflective lighting, load 12–20 blastocysts into the center of the M2 drop using a standard embryo-handling pipette (*see* **Note 2**).
3. Secure the NSET device onto a P-2 pipette that has been set to 1.8 μl.
4. Press the pipette plunger to the first stop, lower the catheter tip into media, and slowly pull the embryos into the tip of the NSET device (*see* **Note 3**). Remove the NSET device tip from the drop.
5. Carefully set the pipette volume to 2.0 μl to create a small air bubble at the catheter tip. Gently lay the pipette with embryos loaded in the catheter on its side. Avoid touching the catheter to any surface prior to the transfer procedure.
6. Place the unanesthetized recipient female on the top of a cage with a wire rack, allowing the mouse to "grab" the cage bar surface. Grasp the mouse near the base of the tail using thumb and forefinger, and angle the tail upward while stabilizing the animal on both sides at the hind legs as shown in Fig. 1 (*see* **Note 4**).
7. Gently place the smaller speculum into the vagina (*see* **Note 5**).
8. Optional: Remove small speculum and replace with the large speculum. If desired, use an adequate light source and visualize the cervix. The transfer procedure can also be performed with the larger speculum, if preferred.

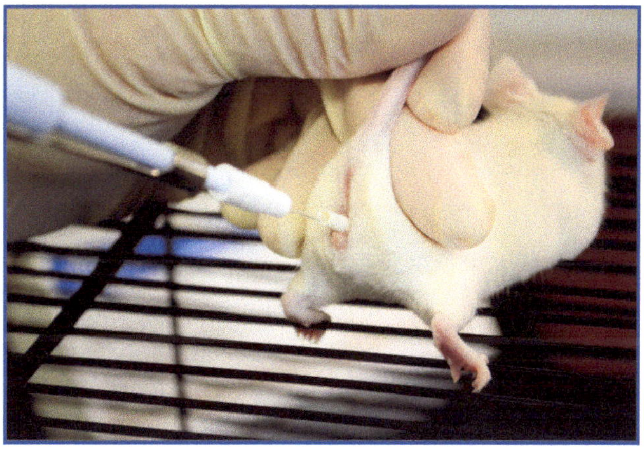

Fig. 1 Nonsurgical embryo transfer procedure. A 2.5 dpc pseudopregnant female CD1 mouse recipient is positioned for NSET device insertion. The mouse is resting on a wire cage top and is stabilized at the tail and at both sides on the front of the hind legs. The small speculum is inserted vaginally and acts as a guide and a stop for the device

9. While holding the female mouse with one hand as described, carefully pick up the pipette and insert the tip of the NSET device into the speculum and through the cervix (*see* **Note 6**).
10. Once the hub of the NSET device contacts the speculum, expel embryos by pressing the pipette plunger (*see* **Note 7**).
11. Gently and slowly remove the NSET device and then remove the speculum (*see* **Note 8**).

4 Notes

1. The lid is preferred as it has a shorter rim and thus is easier to access embryos in the drop. Prepare one drop for each embryo transfer. Do not, at any time, layer mineral oil over the drops of M2 as introducing oil into the uterine horn may negatively affect embryo transfer. Note that the media drops will evaporate over time.
2. The optimal number of embryos to transfer may vary depending upon mouse strain and any manipulations the embryos have received.
3. Visualization of the embryos under a low magnification may aid loading of the embryos into the device.
4. When performed correctly, this handling technique will cause the mouse to hold still for the duration of the procedure. No anesthesia or analgesia is required. Practicing the holding technique and passage of the catheter through the cervix without embryos is highly recommended. CD1 (ICR) mice are widely used as recipients as they are easy to handle and tend to be attentive to their pups.
5. The speculum must be fully inserted at the time of embryo transfer. Once inserted, the speculum can be pushed out by the animal. If this occurs, simply reinsert. The transfer procedure can be performed in most animals using the smaller speculum.
6. When properly positioned, the catheter will glide through the cervix. If the cervix is not entered on the first attempt, gently back the catheter out and attempt to pass through the cervix again. Multiple attempts to find the cervical opening can be made without embryo loss.
7. For a successful embryo transfer, the speculum and NSET device catheter are gently inserted as far as possible. Once the plunger is pressed and the embryos are expelled, it is critical to not release the plunger until the catheter is removed from the uterine horn.
8. Return the mouse to its cage. No post-procedure monitoring is required.

Acknowledgments

Research reported in this publication was funded by the Office of the Director, National Institutes of Health, United States Department of Health and Human Services, under Award Numbers 1R43RR025737-01A1, 2R44RR025737-02, and 8R44OD010958. Additional support was provided by the Kentucky Cabinet for Economic Development, Office of Commercialization and Innovation under the grant agreements KSTC-184-512-11-115 and 184-512-10-096 with the Kentucky Science and Technology Corporation.

References

1. Russell WMS, Burch RL (1959) The principles of humane experimental technique. Methuen & Co., London
2. Green M, Bass S, Spear B (2009) A device for the simple and rapid transcervical transfer of mouse embryos eliminates the need for surgery and potential post-operative complications. BioTechniques 47(5):919–924
3. Steele KH, Hester JM, Stone BJ, Carrico KM, Spear BT, Fath-Goodin A (2013) Nonsurgical embryo transfer device compared with surgery for embryo transfer in mice. J Am Assoc Lab Anim Sci 52(1):17–21
4. Mainigi MA, Olalere D, Burd I, Sapienza C, Bartolomei M, Coutifaris C (2014) Peri-implantation hormonal milieu: elucidating mechanisms of abnormal placentation and fetal growth. Biol Reprod 90(2):26
5. de Waal E, Mak W, Calhoun S, Stein P, Ord T, Krapp C, Coutifaris C, Schultz RM, Bartolomei MS (2014) In vitro culture increases the frequency of stochastic epigenetic errors at imprinted genes in placental tissues from mouse Concepti produced through assisted reproductive technologies. Biol Reprod 90 (2):22
6. Tian X, Anthony K, Neuberger T, Diaz FJ (2014) Preconception zinc deficiency disrupts postimplantation fetal and placental development in mice. Biol Reprod 90(4):83
7. Kaufman MR, Albers RE, Keoni C, Kulkarni-Datar K, Natale DR, Brown TL (2014) Important aspects of placental-specific gene transfer. Theriogenology 82(7):1043–1048
8. Jimenez R, Melo EO, Davydenko O, Ma J, Mainigi M, Franke V, Schultz RM (2015) Maternal SIN3A regulates reprogramming of gene expression during mouse preimplantation development. Biol Reprod 93(4):89
9. Navarrete FA, Alvau A, Lee HC, Levin LR, Buck J, Martin-De Leon P, Santi CM, Krapf D, Mager J, Fissore RA, Salicioni AM, Darszon A, Visconti PE (2016) Transient exposure to calcium ionophore enables *in vitro* fertilization in sterile mouse models. Sci Rep 6:33589
10. Karimi H, Mahdavi P, Fakhari S, Faryabi MR, Esmaeili P, Banafshi O, Mohammadi E, Fathi F, Mokarizadeh A (2017) Altered helper T cell-mediated immune responses in male mice conceived through in vitro fertilization. Reprod Toxicol 69:196–203
11. Woodford C (2011) Use of a non-surgical embryo transfer (NSET) device as an alternative to rodent surgical embryo transfer (ET) and caesarian re-derivation. Anim Technol Welfare 10(1):42–43
12. Ali RB, van der Ahé F, Braumuller TM, Pritchard C, Krimpenfort P, Berns A, Huijbers IJ (2014) Improved pregnancy and birth rates with routine application of nonsurgical embryo transfer. Trans Res 23(4):691–695
13. Moreno-Moya JM, Ramírez L, Vilella F, Martínez S, Quinonero A, Noguera I, Pellicer A, Simon C (2014) Complete method to obtain, culture, and transfer mouse blastocysts nonsurgically to study implantation and development. Fertil Steril 101(3):e13

Chapter 10

Nuclear Transfer and Cloning

Ling Liu

Abstract

Nuclear transfer (NT) and cloning offers a unique and powerful experimental tool to study the mechanisms of gene reprogramming, to establish embryonic stem cells from somatic cells, and to clone live offspring. The process of NT involves two different cells. The first are oocytes, which are enucleated and provide the cellular components required for gene reprogramming and early embryo development. The second are donor cells, and their nuclei are injected into the enucleated oocytes. The donor cell nuclei are reprogrammed in the oocyte cytoplasm. The reconstructed oocytes are then activated and will begin to cleave. The embryos now have a genome identical to the original donor cells. These embryos can be used for basic research or implanting into foster mothers to develop to term. The protocols for mouse NT were developed in 1998 with assistance of the piezo drill device, and in 2000 without assistance of the piezo drill device. In this chapter, a comprehensive update on the techniques and protocols employed to generate mice by NT cloning are summarized based on the latest publications.

Key words Nuclear transfer, Cloning, Somatic cells, Oocytes, Oocyte activation, Embryo transfer, Mouse

1 Introduction

In mammals, nuclear transfer (NT) has been well known as "cloning" since Dolly was reported in 1997 [1]. Since then, animals that have been cloned by NT include sheep [1–3], cattle [4–7], goats [8–10], pigs [11, 12], dogs [13–15], cats [16], rabbit [17], and mice [18–20]. The word "cloning" has caused past misunderstanding of the nature of NT. Cloning was misunderstood as to produce a group of offspring who are the same as the donor both genetically and phenotypically. In actuality, the individuals produced by NT are genetically identical, but they may not be identical phenotypically; for example, they may have different colors and/or different color patterns. NT is a process to only create embryos with original nuclei replaced by donor nuclei.

It is well known that when a sperm fertilizes an oocyte, an embryo forms. That is where new life begins. It is also well established that both sperm and oocytes are germ cells. These are not

Melissa A. Larson (ed.), *Transgenic Mouse: Methods and Protocols*, Methods in Molecular Biology, vol. 2066,
https://doi.org/10.1007/978-1-4939-9837-1_10,

somatic cells, but rather terminally differentiated cells, although both of them have been preprogrammed during the long processes of sperm and oocyte development [21]. As a result, sperm and oocytes have a cell-type-specific memory that have been imposed during differentiation [22]. There are two major points we can learn from this natural process of fertilization. First, sperm DNA is reprogrammed and/or the cell-type-specific memory is erased during fertilization. Second, the stage of the oocyte at fertilization is at metaphase II (MII). It is clear that the MII oocyte contains all of the non-nuclear cellular components required for the early development of an embryo and is effectively the sole contributor to the reprogramming activities. This is why MII oocytes are almost extensively used for the preparation of cytoplasm for NT.

The NT technique is a straightforward one. The nuclear material is removed from the MII oocyte, and the donor cell nucleus is transferred into the enucleated oocyte-cytoplast. Oocytes can be either obtained from naturally ovulated or superovulated animals, or they are collected as immature ova from ovarian follicles and then matured in vitro. It has been demonstrated that oocytes matured in vitro are not of the same developmental potential as naturally-matured ones [23].

Indeed, a wide array of abnormalities and defects have been reported in NT-cloned animals, both before and after birth [4, 18, 19, 24–30]. The most notable defects are increased birth size, placenta defects, and lung, liver, joint, brain, kidney, and cardiovascular pathologies [27, 31]. Multiple failures in several aspects of development are likely to contribute to the defects reported in NT-cloned animals, as no single cause is responsible for all the pathologies. It has been thought to be especially problematic in two processes, DNA reprogramming and gene imprinting [32–36]. To date, the process of donor cell nuclear reprogramming and imprinting remains complex and very poorly understood.

Despite these issues, NT cloning offers the unique opportunity to obtain totipotent cells from fully differentiated cells. Therefore, NT is a new and attractive experimental model for studying the mechanisms of cell differentiation and dedifferentiation. NT cloning is also a method to produce individuals genetically identical to an adult and in some cases facilitates transgenesis. In addition, embryonic stem cells can be obtained by NT cloning and organ stem cells can be derived from embryonic stem cells. Hence, theoretically, NT cloning could contribute to cell therapy by autografting. NT has a large range of multidisciplinary applications, such as basic research, transgenesis, animal reproduction, human reproduction, cell therapy, xenografting, pharmaceutical production, as well as conservation of invaluable genetic resources [22, 37–43].

Mice are the most common experimental animals in laboratory research of biology and medicine. A stable NT cloning method in the mouse was developed in 1989 in Honolulu, which was referred to as the "Honolulu Method" [44] with the piezo drill assistance and then a simplified method was published in 2000 [45] without the piezo assistance. In this chapter, a comprehensive update on the techniques and protocols employed to generate mice by NT cloning are summarized based on the latest publications.

2 Materials

2.1 MII Oocytes

1. Pregnant mare's serum gonadotrophin (PMSG).
2. Human chorionic gonadotrophin (hCG).
3. M2 with hyaluronidase.
4. Culture dishes (35 mm, and/or 60 mm) and Nunc 4-well plates.
5. Pipettes and Pipette tips (1000, 200, 10 μl).
6. Operation medium: M2 medium or CZB medium. 20 mM HEPES should be added and $NaHCO_3$ should be decreased to 4.15 mM if M2 medium is made in-house. Cytochalasin B or D (2–5 μg/ml) is added for enucleation.
7. For practical purposes, the most commonly used strains for successful NT cloning for oocyte donors are hybrids, such as B6D2F1 and B6C3F1. The best birth rate was also obtained with the 129 strain, followed by the DBA/2 strain [22]. Females at 6–12 weeks of age give less numbers of oocytes, but they are of a higher quality, compared to the females at 3–4 weeks of age.

2.2 Nuclear Donor Cells

Somatic cell donors should be selected based on your research interests. They may be any mouse strain. Cumulus cells with a B6D2F1 genetic background have been the standard nuclear donor source and have been used as controls for the assessment of other donor cells. The 129 strain gives higher birth rates followed by the DBA/2 strain. Embryonic stem cells have been reported to support higher efficiency in cloning mice. Of particular difficulty to clone are cells of neural lineage cells, T and B cells [22]).

2.3 Enucleation

1. Mineral Oil.
2. Cytochalasin B or D.
3. Holding pipette: Outer diameter 80–100 μm, inner diameter 10–15 μm.
4. Enucleation pipette: Inner diameter 8–10 μm blunt pipette.
5. PiezoDrill: PMM, PMAS-CT150 or Burleigh PiezoDrill or Eppendorf PiezoXpert.

6. Micromanipulator: Narishige (NT-88-V3) or Eppendorf (Transferman NK2) system.
7. Inverted microscope with Hoffman optics (Olympus 1X71 or Nikon Eclipse Ti).

2.4 Nuclear Transfer

1. Polyvinylpyrrolidone (PVP) (*see* **Note 1**).
2. Washing medium: M2, 12% PVP.
3. Fluorinert FC-70.
4. Injection pipette: Inner diameter 6–7 μm blunt pipette (varies upon the size of donor nucleus) (*see* **Note 2**).

2.5 Activation, Culture, Embryo Transfer

1. Activation medium: 10 mM SrCl2 in calcium-free culture medium. Cytochalasin B or D: 2–5 μg/ml is added to prevent extrusion of polar body. Trichostatin (TSA): 5 nM is added to promote histone acetylation. Media with BSA are stored at 4 °C for no more than 3 weeks. Medium for embryo culture is equilibrated at 37 °C overnight or at least 2–3 h in the incubator with 5% CO_2 and saturated humidity. Medium for operation is warmed up to room temperature before use. Medium compositions are indicated in Table 1. All media are commercially available and can be purchased.
2. CO_2 incubator.
3. Mice used for surrogate mothers for embryo transfer can be any strain. One of the most widely used strains is ICR (CD1). C57/B6 and hybrids are also used in some laboratories. Females serving as foster mothers should be older than 6–8 weeks.

3 Methods (*See* Note 3)

3.1 Collection of MII Oocytes

1. Mice should be fed ad libitum with a standard diet and maintained in temperature- and light-controlled rooms. Experiments must comply with The Care and Use of Laboratory Animals and all related regulations.
2. Superovulation can be induced by injection of 5 IU of PMSG, and 48 h later, followed by 5 IU hCG.
3. In vivo-matured MII cumulus–oocyte-complexes are collected into the operation medium from the oviducts of the oocyte donor mice 12–13 h after hCG injection.
4. Cumulus cells are removed by exposure to 0.1% hyaluronidase and pipetting gently up and down for several times with 200 μl pipette until cumulus cells are released.
5. MII oocytes are washed twice with operation medium and held in culture medium until required.

Table 1
Composition of media (mM)

Component	MW	M2	CZB	KSOM	M 16
NaCl	58.44	94.66	81.62	95	94.6
KCl	75	4.78	4.83	2.5	4.75
KH_2PO_4	136	1.19	1.18	0.35	1.19
$MgSO_4.7H_2O$	246	1.19	1.18	0.2	1.19
$CaCl_2.2H_2O$	147	1.71	1.71	1.71	1.71
$NaHCO_3$	84	4.15	25	25	25
Na lactate (60%)	112.06	23.28	31.3	10	23.28
Na pyruvate	110	0.33	0.27	0.2	0.33
Glucose	180	5.56	0 (5.56)	0.2	5.56
L-Glutamine or Alanyl-glutamine	146.14, 217.22		1	1	
EDTA	292.24		0.11	0.01	
HEPES	238	20.85			
BSA (mg/ml)		4	5	1	4
AAs 50×				0.5×	
Non AAs 100×				0.5×	
Penicillin G (g/L) (IU/ml)		0.06 (100)	0.06 (100)	0.06 (100)	0.06
Streptomycin (g/L)		0.05	0.05	0.05	0.05
Phenol red (ml)		10	10	10	10
Reference		Whittingham [46]	Chatot et al. [47]	Lawitts and Biggers [48] Summers et al. [49]	Whittingham [46]

3.2 Nuclear Donor Cell Preparation

1. Nuclear donor cell preparation depends on the types of cells you are using. Standard cell culture and handling skills required. Donor cells should be isolated into individual cells and washed 3–4 times with operation medium and held there until use. Cumulus cells are easier to prepare. Cells can be collected when denuding oocytes. Fibroblasts have to be prepared days before NT without passaging. ES cells have to be established before NT.
2. All cells have to be suspended in high concentration (1–2 million/ml).

Fig. 1 Layout of the enucleation dish

3.3 Enucleation

1. Group of MII oocytes (10–15) are transferred into a droplet of operation medium containing 5 µg/ml cytochalasin B or D, which is added previously. The layout of the droplets is shown in Fig. 1. All microdrops are covered with mineral oil.
2. Oocytes are held with a holding pipette and rotated until the metaphase spindle is located opposite of the holding pipette, 2–4′ or 8–10′ clock position.
3. Locate the enucleation pipette and make a hole in the zona pellucida (ZP) by applying several piezo pulses (*see* **Note 4**). Adjust the power and speed correctly so that the ZP can be cut easily.
4. Push the enucleation pipette forward through the hole. The metaphase chromosome spindle complex, visible as a translucent region in the cytoplasm, is aspirated into the pipette with minimal volume of oocyte cytoplasm and removed gently.
5. Enucleation for each group should be done within 15–20 min.
6. All oocytes in one group are transferred into droplets of culture medium and held there for up to 2 h.

3.4 Nuclear Transfer

1. Preparation of injection pipette: The injection pipette is loaded with oil and then backloaded with a small amount of FC-70 (3–5 mm column) (*see* **Note 5**). Test in PVP solution droplet and make sure you have a smooth control of the flow.
2. Isolation of donor nuclei: Donor cells are added into the droplets of washing medium (1–3 µl). Nuclei are isolated by gently aspirating the cell in and out of the injection pipette until the nuclei are devoid of visible cytoplasmic material. Piezo

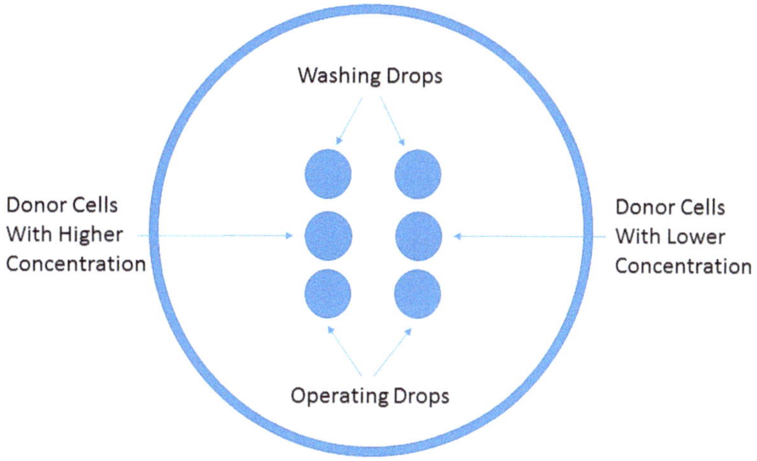

Fig. 2 Layout of the nuclear transfer dish

pulses may be applied if needed. Load 3–4 nuclei into the injection pipette at once.

3. Injection of donor nuclei: Enucleated oocytes are placed into a droplet of operation medium in groups of 10–20. Each nucleus is injected into a enucleated oocyte separately within 4–5 min of its isolation. The injection of donor cell nuclei should be done within 15–20 min for each group. Layout of the droplets is shown in Fig. 2.
   (a) Push one nucleus forward near to the tip of the pipette and advance the pipette to the opposite side of the oocyte.
   (b) One low-power piezo pulse is applied to break the membrane. Expel the donor nucleus into the enucleated oocyte cytoplasm.
   (c) Gently withdraw the injection pipette out of the oocyte.
4. Washing injection pipette: The injection pipette can be washed in the washing medium if it gets stuck or dirty. Application of several piezo pulses will help the cleaning process.
5. Recovery of injected oocytes: Injected oocytes are kept in the operation medium at room temperature for at least 15–20 min and then moved into culture droplets. Oocytes are cultured for at least 30 min at 37 °C before activation (*see* **Note 6**).

3.5 Oocyte Activation and Culture

1. At 1–3 h after nuclear injection, the oocytes are activated by incubation in the embryo culture medium supplemented with 10 mM SrCl2 for 6 h.
2. To prevent polar body extrusion, 2–5 μg/ml cytochalasin B or D may be added in the activation medium.
3. To promote histone deacetylation, 5 nM TSA may be added.

4. At the end of the activation treatment, the oocytes are washed through three times with fresh culture medium to remove the SrCl2 and cultured in the culture medium for overnight.
5. Pronuclear formation can be checked at the end of activation.
6. Two-cell stage embryos can be obtained after overnight culture.
7. These two-cell stage embryos can be used for fresh embryo transfer immediately or cryopreserved for later use.
8. Two-cell embryos can also be continually cultured to the blastocyst stage for fresh embryo transfer or cryopreservation (*see* **Note 7**).

3.6 Embryo Transfer

1. Mate the estrous ICR females with proven sterile ICR males at 1:1 or 2:1 on the same day of NT.
2. Check plugs in the early morning of the next day. These plugged females serve as pseudopregnant mothers.
3. A group of 15–20 two-cell stage embryos are transferred into the oviducts of the plugged females at 0.5 day post-coitus (dpc).

Embryo Transfer (ET)

Animal	DPC	Embryo Stage	ET Site
Mouse	0.5	1-8 cell	Oviduct
	2.5	Morula, Blastocyst	Uterine Horn

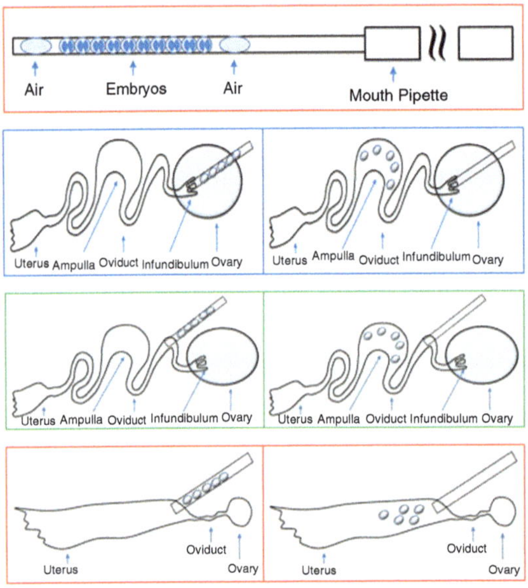

1. Inject animals with appropriate of anesthesia (Ketamine)
2. Protect eyes with Puralube from drying
3. Shave the lower back, wipe with iodine and then with 70% alcohol
4. Make a dorsal incision in the skin (0.5-1cm) and then in the body wall
5. Carefully grasp the fat pad with ovary through the body wall
6. Clamp the fat pad to expose the ampula and secure it
7. Load embryos into the pipette regardless the ET site)
8. Transfer embryos into oviducts or uterine horns
9. Unclamp the fat pad and place the fat pad ovary, oviduct and uterine horn back into the body cavity
10. Close the skin incision with tissue adhesive
11. Administer Banamine
12. Place the animal on a heating pad until it recovers from anesthesia
13. Fill out the Embryology Post Operative Report
14. Between animals, instruments must be wiped clean with 70% alcohol

Fig. 3 Embryo transfer

4. A group of 10–15 blastocyst stage embryos are transferred into the uterine horns of the plugged females at 2.5 dpc (Fig. 3).
5. Pups can be delivered naturally or by caesarean section. Caesarean section is performed between 18.5 dpc and 20.5 dpc (*see* **Note 8**).
6. Lactating foster mothers are used to raise these live pups by mixing the cloned pups with the bedding material of the lactating foster mother's cage and removing some pups from the foster mother's litter.
7. Label or mark the cloned pups properly.
8. Tissue samples from cloned pups (tail), foster mother's (tail), foster mother's pups (tail), and donor (tail or cells) are collected for genetic tests.

4 Notes

1. The quality of PVP varies between bottles even when using the same batch. Quality control of all the media and laboratory systems should be conducted regularly or when new reagents are used.
2. It is very critical to choose the right pipettes for oocyte enucleation and nuclear injection. The inner diameter of injection pipettes should be chosen based on the donor cell type, the smaller the better.
3. NT cloning is a series of steps. The workflow and outline of procedures, including timing, should be managed reasonably.
4. If no PiezoDrill is equipped, oocyte enucleation and nuclear injection can be performed as reported by Zhou [45].
   (a) The outer diameter of the enucleation pipette is not larger than 10 μm.
   (b) The tip of the enucleation pipette has to be beveled and spiked.
   (c) Push the enucleation pipette through the ZP and adjust the tip to maintain close contact with the cytoplasmic membrane above the spindle area.
   (d) Apply negative pressure to draw the cytoplasm with the spindle into the enucleation pipette and pull the pipette slowly out of the ZP. A small amount of karyoplast with spindle can be isolated and removed.
   (e) Donor nuclei can be isolated by gentle aspiration in and out of the injection pipette until virtually no cytoplasm surrounds them.

(f) The nucleus is pushed to the tip of injection pipette, and the pipette is advanced gently through the hole made by enucleation.

(g) Push the pipette gently through the ooplasm until the tip almost reaches the opposite side of the oocyte.

(h) A slight negative pressure is applied to aspirate some of the cytoplasm into the pipette until the cytoplasmic membrane is ruptured.

(i) Positive pressure is applied following the rupture of the cytoplasmic membrane.

(j) Expel the donor nucleus together with the surrounding cytoplasm into the oocyte.

(k) Withdraw the injection pipette carefully.

5. Backload of FC-70 into the injection pipette is important to keep the pipette heavier and give the operator better control. If mercury is allowed in your laboratory, FC-70 can be replaced with mercury for a much better energy transfer.
6. Injected oocytes have to be maintained in room temperature for at least 15–20 min to allow the oocyte membranes to recover.
7. Embryos at the two-cell stage can be frozen or transferred into recipients freshly. If you prefer non-surgical embryo transfer, then embryos should be cultured to the blastocyst stage before being transferred.
8. Caesarean section is recommended for delivery of cloned mice to avoid pup death because of abnormalities in the placenta.

Acknowledgements

The author would like to thank supervisors, Mrs. Jufeng Qian, Northwest Agriculture University; Dr. Mark Westhusin, Texas A&M University; Dr. Colin McKerlie, Toronto Centre for Phenogenomics; and Mr. Steven Sansing, Charles River Laboratory.

References

1. Wilmut I, Schnieke AE, McWhir J, Kind AJ, Campbell KH (1997) Viable offspring derived from fetal and adult mammalian cells. Nature 385(6619):810–813
2. McCreath KJ, Howcroft J, Campbell KH, Colman A, Schnieke AE, Kind AJ (2000) Production of gene-targeted sheep by nuclear transfer from cultured somatic cells. Nature 405(6790):1066–1069
3. Loi P, Ptak G, Barboni B, Fulka J Jr, Cappai P, Clinton M (2001) Genetic rescue of an endangered mammal by cross-species nuclear transfer using post-mortem somatic cells. Nat Biotechnol 19(10):962–964
4. Cibelli JB, Stice SL, Golueke PJ, Kane JJ, Jerry J, Blackwell C, Ponce de Leon FA, Robl JM (1998) Cloned transgenic calves produced from nonquiescent fetal fibroblasts. Science 280(5367):1256–1258

5. Kato Y, Tani T, Sotomaru Y, Kurokawa K, Kato J, Doguchi H, Yasue H, Tsunoda Y (1998) Eight calves cloned from somatic cells of a single adult. Science 282 (5396):2095–2098
6. Kubota C, Yamakuchi H, Todoroki J, Mizoshita K, Tabara N, Barber M, Yang X (2000) Six cloned calves produced from adult fibroblast cells after long-term culture. Proc Natl Acad Sci U S A 97(3):990–995
7. Lanza RP, Cibelli JB, Blackwell C, Cristofalo VJ, Francis MK, Baerlocher GM, Mak J, Schertzer M, Chavez EA, Sawyer N, Lansdorp PM, West MD (2000) Extension of cell life-span and telomere length in animals cloned from senescent somatic cells. Science 288 (5466):665–669
8. Baguisi A, Behboodi E, Melican DT, Pollock JS, Destrempes MM, Cammuso C, Williams JL, Nims SD, Porter CA, Midura P, Palacios MJ, Ayres SL, Denniston RS, Hayes ML, Ziomek CA, Meade HM, Godke RA, Gavin WG, Overstrom EW, Echelard Y (1999) Production of goats by somatic cell nuclear transfer. Nat Biotechnol 17(5):456–461
9. Keefer CL, Baldassarre H, Keyston R, Wang B, Bhatia B, Bilodeau AS, Zhou JF, Leduc M, Downey BR, Lazaris A, Karatzas CN (2001) Generation of dwarf goat (Capra hircus) clones following nuclear transfer with transfected and nontransfected fetal fibroblasts and in vitro-matured oocytes. Biol Reprod 64(3):849–856
10. Keefer CL, Keyston R, Lazaris A, Bhatia B, Begin I, Bilodeau AS, Zhou FJ, Kafidi N, Wang B, Baldassarre H, Karatzas CN (2002) Production of cloned goats after nuclear transfer using adult somatic cells. Biol Reprod 66 (1):199–203
11. Polejaeva IA, Chen SH, Vaught TD, Page RL, Mullins J, Ball S, Dai Y, Boone J, Walker S, Ayares DL, Colman A, Campbell KH (2000) Cloned pigs produced by nuclear transfer from adult somatic cells. Nature 407(6800):86–90
12. Onishi A (2002) Cloning of pigs from somatic cells and its prospects. Cloning Stem Cells 4 (3):253–259
13. Lee BC, Kim MK, Jang G, Oh HJ, Yuda F, Kim HJ, Hossein MS, Kim JJ, Kang SK, Schatten G, Hwang WS (2005) Dogs cloned from adult somatic cells. Nature 436(7051):641
14. Hossein MS, Jeong YW, Park SW, Kim JJ, Lee E, Ko KH, Hyuk P, Hoon SS, Kim YW, Hyun SH, Shin T, Hwang WS (2009) Birth of beagle dogs by somatic cell nuclear transfer. Anim Reprod Sci 114(4):404–414
15. Hossein MS, Jeong YW, Park SW, Kim JJ, Lee E, Ko KH, Kim HS, Kim YW, Hyun SH, Shin T, Hawthorne L, Hwang WS (2009) Cloning Missy: obtaining multiple offspring of a specific canine genotype by somatic cell nuclear transfer. Cloning Stem Cells 11 (1):123–130
16. Shin T, Kraemer D, Pryor J, Liu L, Rugila J, Howe L, Buck S, Murphy K, Lyons L, Westhusin M (2002) A cat cloned by nuclear transplantation. Nature 415(6874):859
17. Chesne P, Adenot PG, Viglietta C, Baratte M, Boulanger L, Renard JP (2002) Cloned rabbits produced by nuclear transfer from adult somatic cells. Nat Biotechnol 20(4):366–369
18. Wakayama T, Perry AC, Zuccotti M, Johnson KR, Yanagimachi R (1998) Full-term development of mice from enucleated oocytes injected with cumulus cell nuclei. Nature 394 (6691):369–374
19. Wakayama T, Yanagimachi R (1999) Cloning of male mice from adult tail-tip cells. Nat Genet 22(2):127–128
20. Wakayama T, Shinkai Y, Tamashiro KL, Niida H, Blanchard DC, Blanchard RJ, Ogura A, Tanemura K, Tachibana M, Perry AC, Colgan DF, Mombaerts P, Yanagimachi R (2000) Cloning of mice to six generations. Nature 407(6802):318–319
21. Kafri T, Ariel M, Brandeis M, Shemer R, Urven L, McCarrey J, Cedar H, Razin A (1992) Developmental pattern of gene-specific DNA methylation in the mouse embryo and germ line. Genes Dev 6(5):705–714
22. Ogura A, Inoue K, Wakayama T (2013) Recent advancements in cloning by somatic cell nuclear transfer. Philos Trans R Soc Lond Ser B Biol Sci 368(1609):20110329
23. Fulka J Jr, Fulka H (2007) Somatic cell nuclear transfer (SCNT) in mammals: the cytoplast and its reprogramming activities. Adv Exp Med Biol 591:93–102
24. Schnieke AE, Kind AJ, Ritchie WA, Mycock K, Scott AR, Ritchie M, Wilmut I, Colman A, Campbell KH (1997) Human factor IX transgenic sheep produced by transfer of nuclei from transfected fetal fibroblasts. Science 278 (5346):2130–2133
25. Renard JP, Chastant S, Chesne P, Richard C, Marchal J, Cordonnier N, Chavatte P, Vignon X (1999) Lymphoid hypoplasia and somatic cloning. Lancet 353(9163):1489–1491
26. Wells DN, Misica PM, Tervit HR (1999) Production of cloned calves following nuclear transfer with cultured adult mural granulosa cells. Biol Reprod 60(4):996–1005
27. Hill JR, Burghardt RC, Jones K, Long CR, Looney CR, Shin T, Spencer TE, Thompson JA, Winger QA, Westhusin ME (2000)

Evidence for placental abnormality as the major cause of mortality in first-trimester somatic cell cloned bovine fetuses. Biol Reprod 63 (6):1787–1794
28. De Sousa PA, King T, Harkness L, Young LE, Walker SK, Wilmut I (2001) Evaluation of gestational deficiencies in cloned sheep fetuses and placentae. Biol Reprod 65(1):23–30
29. Hill JR, Winger QA, Burghardt RC, Westhusin ME (2001) Bovine nuclear transfer embryo development using cells derived from a cloned fetus. Anim Reprod Sci 67(1–2):17–26
30. Ono Y, Shimozawa N, Ito M, Kono T (2001) Cloned mice from fetal fibroblast cells arrested at metaphase by a serial nuclear transfer. Biol Reprod 64(1):44–50
31. Wilmut I (2006) Are there any normal clones? Methods Mol Biol 348:307–318
32. Jaenisch R, Eggan K, Humphreys D, Rideout W, Hochedlinger K (2002) Nuclear cloning, stem cells, and genomic reprogramming. Cloning Stem Cells 4(4):389–396
33. Jaenisch R (2009) Stem cells, pluripotency and nuclear reprogramming. J Thromb Haemost 7 (Suppl 1):21–23
34. Colman A, Burley J (2014) Human somatic cell reprogramming: does the egg know best? Cell Stem Cell 15(5):531–532
35. Tiemann U, Wu G, Marthaler AG, Scholer HR, Tapia N (2016) Epigenetic aberrations are not specific to transcription factor-mediated reprogramming. Stem Cell Reports 6 (1):35–43
36. Wan Y, Deng M, Zhang G, Ren C, Zhang H, Zhang Y, Wang L, Wang F (2016) Abnormal expression of DNA methyltransferases and genomic imprinting in cloned goat fibroblasts. Cell Biol Int 40(1):74–82
37. Houdebine LM (2000) Transgenic animal bioreactors. Transgenic Res 9(4–5):305–320
38. Houdebine LM (2005) Use of transgenic animals to improve human health and animal production. Reprod Domest Anim 40 (4):269–281
39. Houdebine LM (2009) Applications of genetically modified animals. J Soc Biol 203 (4):323–328
40. Lewandowski J, Kurpisz M (2016) Techniques of human embryonic stem cell and induced pluripotent stem cell derivation. Arch Immunol Ther Exp 64(5):349–370
41. Niemann H, Petersen B (2016) The production of multi-transgenic pigs: update and perspectives for xenotransplantation. Transgenic Res 25(3):361–374
42. Rogers CS (2016) Genetically engineered livestock for biomedical models. Transgenic Res 25(3):345–359
43. Su F, Wang Y, Liu G, Ru K, Liu X, Yu Y, Liu J, Wu Y, Quan F, Guo Z, Zhang Y (2016) Generation of transgenic cattle expressing human beta-defensin 3 as an approach to reducing susceptibility to Mycobacterium bovis infection. FEBS J 283(5):776–790
44. Kishigami S, Wakayama S, Thuan NV, Ohta H, Mizutani E, Hikichi T, Bui HT, Balbach S, Ogura A, Boiani M, Wakayama T (2006) Production of cloned mice by somatic cell nuclear transfer. Nat Protoc 1(1):125–138
45. Zhou Q, Boulanger L, Renard JP (2000) A simplified method for the reconstruction of fully competent mouse zygotes from adult somatic donor nuclei. Cloning 2(1):35–44
46. Whittingham DG (1971) Culture of mouse ova. J Reprod Fertil Suppl 14:7–21
47. Chatot CL, Ziomek CA, Bavister BD, Lewis JL, Torres I (1989) An improved culture medium supports development of random-bred 1-cell mouse embryos in vitro. J Reprod Fertil 86(2):679–688
48. Lawitts JA, Biggers JD (1993) Culture of preimplantation embryos. Methods Enzymol 225:153–164
49. Summers MC, McGinnis LK, Lawitts JA, Raffin M, Biggers JD (2000) IVF of mouse ova in a simplex optimized medium supplemented with amino acids. Hum Reprod 15 (8):1791–1801

Chapter 11

Sperm-Mediated Genetic Modifications

Marialuisa Lavitrano, Laura Farina, Maria Grazia Cerrito, and Roberto Giovannoni

Abstract

The ability to introduce controlled modifications of the genome of animals represents an important tool for biomedical and veterinary research. Among transgenic techniques, we describe here the sperm-mediated gene transfer method that is based on the spontaneous ability of sperm cells to bind and internalize exogenous DNA and to carry it to the oocyte during fertilization, producing genetically modified animals.

Key words Spermatozoa, Transgenic animals, SMGT, Transgenesis, Mouse

1 Introduction

The ability of introducing controlled modifications of the genome of animals represented, since the firstly reported transgenic mouse [1], an important tool for biomedical and veterinary research [2, 3]. A transgenic animal is defined as one with transmissible genetic modification(s) induced by transgenesis experiments [4], and over the years, many approaches and techniques for the production of genetically modified animals have been developed and optimized. In the early 1980s, one of the first to be shown effective in mammals and still the most common and widely used across species is pronuclear microinjection [5, 6]. This procedure, used to overexpress or underexpress certain genes or to express genes entirely new to the host organism, involves the direct microinjection of DNA molecules into the male pronucleus of a zygote. Usually, the transgene is a eukaryotic expression cassette encoding for a single gene or a combination of genes from the same or another species, that randomly integrates into the mouse genome. In 1985, Smithies and collaborators were able to substitute endogenous beta-globin allele with a mutated one in a mammalian cell line [7] and this led to the development of the embryonic stem cell transfer as transgenesis technique [8]. Other procedures such as somatic or embryonic cell nuclear transfer and viral-based

Melissa A. Larson (ed.), *Transgenic Mouse: Methods and Protocols*, Methods in Molecular Biology, vol. 2066, https://doi.org/10.1007/978-1-4939-9837-1_11, © Springer Science+Business Media, LLC, part of Springer Nature 2020 constructs as vectors for the introduction of exogenous DNA into embryos have also been developed [6] and are discussed in other chapters of this book.

The capacity of sperm cells to capture foreign DNA had been reported in the pioneer study of Brackett et al. [9]. The sperm-mediated gene transfer (SMGT) method developed by Lavitrano and collaborators, is based on the intrinsic ability of sperm cells to bind and internalize exogenous DNA and to transfer it into the egg at fertilization [10–14].

The interaction between sperm cells and DNA is not a random event, and it is mediated by DNA-binding proteins (DBP) present on the sperm cell surface. The formation of the exogenous DNA/DBP complex is dependent on MHC class II expression, and it triggers CD4-mediated internalization [12, 15]. The DNA/DBP/CD4 complex reaches the nucleus and a small fraction of the exogenous DNA molecules integrate in unique or multiple sites within the genome [15, 16], while the remaining part is degraded by a sperm endogenous nuclease [17]. The spontaneous internalization of exogenous DNA molecules by sperm cells could affect the species' genome integrity, and there are two natural mechanisms to antagonize this process: (1) the presence of the inhibitory factor IF-1, which is abundant in mammalian seminal fluid and expressed on the sperm membrane in marine animals; and (2) the nuclease activity in spermatozoa or in the seminal fluid (fish) [15, 18]. It has been demonstrated that epididymal and ejaculated spermatozoa uptake DNA molecules only after sequential washing procedures removing seminal fluid [17, 19]. The SMGT protocol has been adapted for use in several species and selection of sperm donors and optimization of DNA uptake protocols have been demonstrated to be key steps in the SMGT procedure [19–24].

We first described the SMGT procedure in the mouse [10], and then we successfully adapted the technique for use in large animals; in fact, it was highly efficient for the generation of human decay accelerating factor (hDAF) transgenic pig lines [16, 19]. In addition, SMGT has been successful in generating pigs carrying multigene transgenes by simultaneously introducing three reporter genes, namely, enhanced green fluorescent protein (EGFP), enhanced blue fluorescent protein (EBFP), and red fluorescent protein [25].

Sperm-mediated gene transfer has also been used for the development of transgenic procedures by making use of different methods of fertilization, such as the intracytoplasmic sperm injection (ICSI) [26, 27] and the recently reported round spermatid nucleus injection (ROSI) [28]. The ICSI technique is based on the observation that membrane-disrupted sperm heads were able to support full development when injected into metaphase II mouse oocytes [29]. Spermatozoa subjected to membrane disruption by Triton

X-100, freeze–thawing, or freeze-drying and incubated with exogenous DNA are able to transfer a transgene to the embryos and to the offspring [26]. Similarly, the ROSI technique is based on the use of spermatids, immature spermatogenic haploid cells [30] that are injected into nonfertilized oocytes after incubation with transgenic DNA [28].

Moreover, in order to explore new possibilities of improving the efficiency of SMGT, different groups introduced modifications to the SMGT protocol, such as the use of magnetic nanoparticles (MNPs) associated with labeled oligonucleotides to increase the efficiency of DNA uptake by spermatozoa [31] or the use of sperm cells infected with lentiviral vectors [32].

2 Materials

2.1 Equipment

In vitro fertilization is very sensitive to environmental conditions. Controlled temperature and sterility are the two key parameters for achieving successful experiments and therefore must be accurately set in the working room. Basic equipment for a cell biology laboratory is required.

1. Laminar flow hood.
2. CO_2 incubator.
3. Inverted microscope.
4. Stereomicroscope equipped with heating stage.
5. Fiber optic lighting system.
6. Heating block.
7. Cell counter.
8. Falcon 3653 dishes for squeezing epididymis, collecting eggs, fertilizations, and embryo cultures.

2.2 In Vitro Fertilization

1. Fertilization medium: NaCl 6.970 g/l, $NaHCO_3$ 2.106 g/l, Glucose 1.000 g/l, KCl 0.201 g/l, $Na_2HPO_4{\cdot}2H_2O$ 0.027 g/l, $CaCl_2{\cdot}2H_2O$ 0.264 g/l, $MgCl_2{\cdot}6H_2O$ 0.102 g/l, Na Pyruvate 0.055 g/l. Osmolarity 275–290. The medium is supplemented with 4 mg/ml bovine serum albumin. Filter all media through 0.22 μm filters and store at 4 °C under sterile conditions for no more than 4 weeks (*see* **Note 1**). The medium used for in vitro fertilization experiments (FM) is based on Whittingham's Tyrode solution [33], from which sodium lactate, penicillin, and streptomycin are omitted; NaH_2PO_4 is replaced by 0.15 mM Na_2HPO_4 and NaCl is increased to 120 mM.
2. Ham's F10 medium is hypoxanthine-free, supplemented with $NaHCO_3$ (2.106 g/l) and 4 mg/ml BSA. Osmolarity should also be between 275 and 290.

3. Silicon oil to cover embryo cultures.

2.3 Mouse Manipulations

1. PMSG and hCG, both at 50 IU/ml.
2. Concentrated 40× Avertin anesthetic stock: prepare by dissolving 10 g. tribromoethanol in 10 ml of tert-Amyl alcohol, store at 4 °C. Prepare the working solution by diluting the stock to 1:40 with physiological solution.
3. All the procedures involving animals must be performed according to the guidelines and normative on the use of animals in scientific research.
   (a) Crl:CD1 (CD-1) and B6D2F1/Crl (BDF1) mice purchased from Charles River.
   (b) CD-1 males are routinely used as sperm donors.
   (c) Vasectomized CD-1, tested for sterility, are mated with CD-1 recipients 8–12 h before implanting the embryos.
   (d) CD-1 recipients are retired females. BDF1 females 3–10 weeks' old are routinely used as egg donors.

3 Methods

3.1 Oocyte Collection

1. Mice are maintained in cycles of 13 h light and 11 h dark.
2. Superovulation is induced following intraperitoneal injection of BDF1 mice with 5 IU of PMSG at 8.30–9.30 pm; 48 h later, 5 additional IU of hCG are injected intraperitoneally.
3. Carry out all the procedures (squeezing of epididymis, fertilization, embryo cultures) in the inner wells of Falcon 3653 dishes containing 1 ml of appropriate medium and overlaid with autoclaved silicon oil. Place 4 ml of BSA-free PBS in the outer wells. This arrangement ensures optimal conditions for both cell and embryo survival. It is recommended to prepare two sets of dishes respectively containing FM and Ham's F10, as described above, 15–24 h before starting the experiment and place them in a 37 °C incubator under 7% CO_2 in air to equilibrate both the temperature and pH. To accurately evaluate the pH of different media (optimum pH is 7–7.5), dispense 3 ml aliquots of each medium into conical tubes and incubated in parallel with the dishes.
4. Collect eggs from 3–10-week-old BDF1 female mice, 13 h after induction of superovulation. Females are sacrificed by cervical dislocation, oviducts are removed, and placed in a PBS-containing dish.
5. Place two oviducts in the outer well of each dish and squeeze out the egg cumulus using a 1-ml syringe.
6. Transfer the cumulus in the inner well containing 1 ml of fertilization medium (FM) supplemented with BSA (4 mg/ml,

pre-equilibrated overnight in the incubator) and overlaid with silicon oil.

3.2 Sperm Collection

1. Spermatozoa for in vitro fertilization are obtained from CD-1 male mice. Collect sperm cells from the cauda epididymis of proven males that abstained for at least 3 days but no longer than 1 week. Prepare the sperm suspension by squeezing the terminal part of the *vas deferens* and puncturing the middle part of the epididymis in 1 ml of pre-equilibrated FM supplemented with 4 mg/ml BSA overlaid with silicon oil. Allow the sperm suspension to disperse by incubating the drop for 30 min at 37 °C in 7% CO_2 in air (*see* **Note 2**).
2. After dispersal, count sperm cells as follows: dilute 10 μl of the sperm suspension with 990 μl of HCL 0.1 N. Spot 10 μl in a hemocytometer chamber and count. The concentration of sperm cells is usually 25–36 × 10^6/ml. Withdraw the required amount of sperm cells and mix them with plasmid DNA at a concentration ranging from 0.01–1 μg/10^6 sperm cells. Plasmid DNA can be used either in the linear or in the circular form; plasmids are usually linearized using enzymes that restrict at the cloning sites and separate the gene from the vector backbone sequences.
3. Incubate spermatozoa and DNA for 30 min at 37 °C in 7% CO_2 in the air. The incubation time can be reduced to 15 min without loss of efficiency.

3.3 In Vitro Fertilization and Generating Mice

1. After incubation, withdraw aliquots of 1–2 × 10^6 spermatozoa and add them to the egg-containing dishes—usually 30–70 eggs collected from two oviducts are pooled in one dish. Place dishes containing both sperm cells and eggs in the incubator for 5 h.
2. After incubation, transfer nondegenerated eggs into dishes containing 1 ml of Ham's F10 and incubate overnight (*see* **Note 3**). Assess the fertilization rate 26–28 h after mixing the gametes by measuring the percentage of two-cell embryos (*see* **Note 4**).
3. Transfer two-cell stage embryos [15–20] to a 37 °C pre-warmed dish containing 50 μl of PBS supplemented with 1 mg/ml BSA and overlaid with silicon oil. Under these conditions the dish can temporarily be kept at 37 °C on a heating block outside the incubator.
4. Collect 15–20 embryos in PBS with a micropipette and surgically implant them into the oviduct of plugged retired CD-1 females that had been mated to CD-1 vasectomized males 8–12 h before. During surgical implantation of the embryos,

the foster mothers are anesthetized by intraperitoneal injection of 15 μl of Avertin anesthetic/g of bodyweight.

5. Let the deliveries go to term.
6. Collect biopsies from tail or ear of newborn mice and extract genomic DNA from biopsies to detect the transgenes by standard procedures [34].

4 Notes

1. The source of water for the media preparation is a very critical factor for the efficiency of fertilization. We have found that most commercially available purified H_2O lots are not suitable for fertilization experiments. In our laboratory, water is routinely purified through a Millipore system, repurified through a Barnstead Easypure System containing a Pretreatment cartridge R/O and a Easypure high purity/low TOC Cartridge, and finally quartz-distilled in a Barnstead mega-Pure water still.
2. It is recommended to pool sperm cells squeezed from epididymides of two males.
3. Transfer of fertilized eggs from FM to Ham's F10 takes place outside the incubator; therefore, it is recommended to operate as quickly as possible to minimize embryo damage.
4. Trial fertilization experiments with control epididymal sperm routinely give an efficiency included between 60% and 100%; under optimal fertilization conditions, the use of DNA-loaded sperm cells does not substantially modify this efficiency. In contrast, the efficiency is significantly reduced in those batches of eggs whose fertilization efficiency by control sperm was below 60%: thus, the experiments are not carried on any further below 60%.

Acknowledgments

This work was supported by the Italian Minister of Research and University (FIRB, RBAP06LAHL).

References

1. Gordon JW, Scangos GA, Plotkin DJ et al (1980) Genetic transformation of mouse embryos by microinjection of purified DNA. Proc Natl Acad Sci U S A 77:7380–7384
2. Tan W, Proudfoot C, Lillico SG, Whitelaw CBA (2016) Gene targeting, genome editing: from Dolly to editors. Transgenic Res 25:273–287. https://doi.org/10.1007/s11248-016-9932-x
3. Vandamme TF (2015) Rodent models for human diseases. Eur J Pharmacol 759:84–89. https://doi.org/10.1016/j.ejphar.2015.03.046

4. Th Rülicke, X Montagutelli, B Pintado, R Thon, H J Hedrich, (2016) FELASA guidelines for the production and nomenclature of transgenic rodents. Laboratory Animals 41 (3):301–311
5. Gordon JW, Ruddle FH (1981) Integration and stable germ line transmission of genes injected into mouse pronuclei. Science 214:1244–1246
6. Belizário JE, Akamini P, Wolf P et al (2012) New routes for transgenesis of the mouse. J Appl Genet 53:295–315. https://doi.org/10.1007/s13353-012-0096-y
7. Smithies O, Gregg RG, Boggs SS et al (1985) Insertion of DNA sequences into the human chromosomal beta-globin locus by homologous recombination. Nature 317:230–234
8. Capecchi MR (1989) Altering the genome by homologous recombination. Science 244:1288–1292. https://doi.org/10.1126/science.2660260
9. Brackett BG, Baranska W, Sawicki W, Koprowski H (1971) Uptake of heterologous genome by mammalian spermatozoa and its transfer to ova through fertilization. Proc Natl Acad Sci U S A 68:353–357
10. Lavitrano M, Camaioni A, Fazio VM et al (1989) Sperm cells as vectors for introducing foreign DNA into eggs: genetic transformation of mice. Cell 57:717–723
11. Francolini M, Lavitrano M, Lamia C et al (1993) Evidence for nuclear internalization of exogenous DNA into mammalian sperm cells. Mol Reprod Dev 34:133–139
12. Zani M, Lavitrano M, French D et al (1995) The mechanism of binding of exogenous DNA to sperm cells: factors controlling the DNA uptake. Exp Cell Res 217:57–64. https://doi.org/10.1006/excr.1995.1063
13. Lavitrano M, Maione B, Forte E et al (1997) The interaction of sperm cells with exogenous DNA: a role of CD4 and major histocompatibility complex class II molecules. Exp Cell Res 233(1):58–62. https://doi.org/10.1006/excr.1997.3534
14. Lavitrano M, Busnelli M, Cerrito MG et al (2006) Sperm-mediated gene transfer. Reprod Fertil Dev 18:19–23
15. Magnano A, Giordano R, Moscufo N et al (1998) Sperm/DNA interaction: integration of foreign DNA sequences in the mouse sperm genome. J Reprod Immunol 41:187–196
16. Lavitrano M, Bacci ML, Forni M et al (2002) Efficient production by sperm-mediated gene transfer of human decay accelerating factor (hDAF) transgenic pigs for xenotransplantation. Proc Natl Acad Sci U S A 99:14230–14235. https://doi.org/10.1073/pnas.222550299
17. Maione B, Pittoggi C, Achene L et al (1997) Activation of endogenous nucleases in mature sperm cells upon interaction with exogenous DNA. DNA Cell Biol 16:1087–1097. https://doi.org/10.1089/dna.1997.16.1087
18. Lavitrano M, Giovannoni R, Cerrito MG (2013) Methods for sperm-mediated gene transfer. Methods Mol Biol 927:519–529. https://doi.org/10.1007/978-1-62703-038-0_44
19. Lavitrano M, Forni M, Bacci ML et al (2003) Sperm mediated gene transfer in pig: selection of donor boars and optimization of DNA uptake. Mol Reprod Dev 64:284–291. https://doi.org/10.1002/mrd.10230
20. Cong L, Cao G, Renyu X et al (2011) Reducing blood glucose level in TIDM mice by orally administering the silk glands of transgenic hIGF-I silkworms. Biochem Biophys Res Commun 410:721–725. https://doi.org/10.1016/j.bbrc.2011.05.157
21. Sin FY, Walker SP, Symonds JE et al (2000) Electroporation of salmon sperm for gene transfer: efficiency, reliability, and fate of transgene. Mol Reprod Dev 56:285–288. https://doi.org/10.1002/(SICI)1098-2795(200006)56:2+<285::AID-MRD16>3.0.CO;2-4
22. Campos VF, Amaral MG, Seixas FK et al (2011) Exogenous DNA uptake by South American catfish (*Rhamdia quelen*) spermatozoa after seminal plasma removal. Anim Reprod Sci 126:136–141. https://doi.org/10.1016/j.anireprosci.2011.05.004
23. Collares T, Campos VF, De Leon PM et al (2011) Transgene transmission in chickens by sperm-mediated gene transfer after seminal plasma removal and exogenous DNA treated with dimethylsulfoxide or N, N-dimethylacetamide. J Biosci 36:613–620
24. Wang H, Lin A, Zhang Z, Chen Y (2001) Expression of porcine growth hormone gene in transgenic rabbits as reported by green fluorescent protein. Anim Biotechnol 12:101–110. https://doi.org/10.1081/ABIO-100108336
25. Webster NL, Forni M, Bacci ML et al (2005) Multi-transgenic pigs expressing three fluorescent proteins produced with high efficiency by sperm mediated gene transfer. Mol Reprod Dev 72:68–76. https://doi.org/10.1002/mrd.20316
26. Perry ACF, Wakayama T, Kishikawa H et al (1999) Mammalian transgenesis by intracytoplasmic sperm injection. Science

284:1180–1183. https://doi.org/10.1126/science.284.5417.1180

27. Moisyadi S, Kaminski JM, Yanagimachi R (2009) Use of intracytoplasmic sperm injection (ICSI) to generate transgenic animals. Comp Immunol Microbiol Infect Dis 32:47–60. https://doi.org/10.1016/j.cimid.2008.05.003
28. Moreira P, Pérez-Cerezales S, Laguna R et al (2016) Transgenic mouse offspring generated by ROSI. J Reprod Dev 62:37–42. https://doi.org/10.1262/jrd.2015-105
29. Kimura Y, Yanagimachi R, Kuretake S et al (1998) Analysis of mouse oocyte activation suggests the involvement of sperm perinuclear material. Biol Reprod 58:1407–1415
30. O'Donnell L, O'Bryan MK (2014) Microtubules and spermatogenesis. Semin Cell Dev Biol 30:45–54. https://doi.org/10.1016/j.semcdb.2014.01.003
31. Katebi S, Esmaeili A, Ghaedi K (2016) Static magnetic field reduced exogenous oligonucleotide uptake by spermatozoa using magnetic nanoparticle gene delivery system. J Magn Magn Mater 402:184–189. https://doi.org/10.1016/j.jmmm.2015.11.057
32. Zhang Y, Xi Q, Ding J et al (2012) Production of transgenic pigs mediated by pseudotyped lentivirus and sperm. PLoS One 7:e35335. https://doi.org/10.1371/journal.pone.0035335
33. Whittingham DG (1971) Culture of mouse ova. J Reprod Fertil Suppl 14:7–21
34. Montoliu L (1997) Generation of transgenic mice. A laboratory manual

Chapter 12

Genotyping Genetically Modified (GM) Mice

Neeraj K. Aryal and Jan Parker-Thornburg

Abstract

Prior to generating a new mouse model, it is important to plan the method that will be used to detect which of the mice generated have the mutation(s) desired. Nearly, all types of mutations may be detected using PCR. However, the choice of primers will differ depending upon the method used to generate the model. Transgenic mice should be genotyped across a unique junction fragment. Targeted ES cells used to generate knock-out or knock-in mice should be genotyped using primers from a unique marker in the construct and a region outside of the construct. Targeting in ES cells can also be detected using a genomic Southern blot. Mice targeted using CRISPR/Cas9 should have the region of interest amplified using PCR, and then be assessed for size changes (for large changes in sequence) by Surveyor Assay (for gene knock-out and point mutations) and/or sequenced to verify the mutation. Each of these models has a unique requirement for genotyping, and failure to understand the requirements can easily lead to loss of the gene in subsequent generations.

Key words Genotyping, Genetically engineered mice, Transgenic, Targeted, Knock-out mice, Knock-in mice, CRISPR/Cas9, Germ line transmission, Chimera, Mosaic, PCR

1 Introduction

Genetically engineered mice are now relatively easy to generate. However, once animals are in hand, they need to be tested to determine whether they have the correct genotype. Not all animals produced will contain the introduced gene or mutation. Additionally, even after generations of breeding to establish or maintain a line, not all of the animals in a litter may have the gene of interest unless care was taken to generate a homozygous line. Thus, it is essential that a simple and robust genotyping strategy is developed to assess whether the gene is present or correctly modified PRIOR to generating the mice. Many facilities will require that the genotyping strategy be defined and demonstrated before beginning a project. One should plan ahead, setting up and demonstrating a viable genotyping strategy, to reduce the amount of time needed to assess the animals produced, and thus enable more animal life span for mating to propagate the line.

Melissa A. Larson (ed.), *Transgenic Mouse: Methods and Protocols*, Methods in Molecular Biology, vol. 2066,
https://doi.org/10.1007/978-1-4939-9837-1_12,

The genotyping method of choice is generally dependent upon the specific model to be produced. In all cases, it is essential that the method is specific for detection of the introduced gene. For *transgenic* mice, the DNA introduced into the genome is generally based on a chimeric gene. By designing primers that are specific for the chimeric gene, one can easily use polymerase chain reaction (PCR) for genotyping transgenic mice. *Targeted* mice (whether knock-out or knock-in) have historically been produced by injection of manipulated embryonic stem (ES) cells, where the definitive genotyping is done in the ES cells by long-range PCR or by Southern blot analysis. This genotyping strategy is also used during establishment of the mouse line and then subsequently replaced by a simple PCR once the mouse line is in hand. More recently, targeted mice are produced using *CRISPR/Cas9* or other gene editing nuclease (GEN) methods [1]. While CRISPR/Cas9 models may be generated quickly, they are often quite complex. For genotyping of these models, the initial genotyping usually involves PCR of the region targeted, followed by specialized assays and/or sequencing to determine the various mutations obtained, as mosaicism in the G0 founders can complicate this process [2].

1.1 Transgenic Mice

Transgenic mice are generated by pronuclear injection of a DNA fragment into a fertilized egg. For standard transgenic mouse generation, DNA incorporation upon pronuclear injection is random. Multiple copies of the transgene can be incorporated at the same locus, and on rare occasions, the transgene can be inserted into more than one place in the genome [3]. It is essential to consider these caveats when devising a genotyping strategy. As incorporation is random, one cannot use endogenous sequences as part of the targeting strategy. More importantly, one should NOT base genotyping on any part of the transgene that is also present in the mouse genome, as one will not be able to distinguish between the endogenous gene and the transgene.

Nearly all transgenic mice utilize a chimeric gene, perhaps expressing a mutant structural gene, or a reporter gene that is driven by a tissue or stage-specific promoter [4]. One could genotype for the promoter; however, the endogenous gene will be present, confounding the analysis. One could genotype for the mutant gene or the reporter gene, but this may not be informative if there is more than one insertion site, and only one site is expressed at the proper levels or if there are permutations of the gene that are in the genome (e.g., genotyping for Cre recombinase when there could be different tissue-specific promoters driving Cre in the genome). The most informative genotyping design is to develop primers that span a junction within the chimeric gene or that incorporate the mutation being introduced (Fig. 1). By doing so, one uses primers that can detect only the transgene.

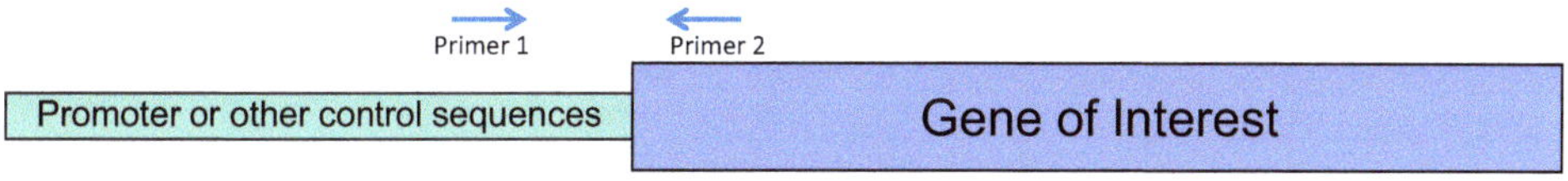

Fig. 1 Selecting primers that cross a junction for detection of **transgene** insertion

1.2 Gene-Targeted Mice Using ES Cell Technology

Targeting in ES cells is dependent upon determining whether the correct deletion or insertion is present in the resident gene [5]. Specific requirements to assay for the DNA modification may be suggested by your targeting facility. Typically, this involves devising a scheme to determine the correct targeting of the clone(s). The strategy has to be rigorously tested and proven to be both reliable and reproducible. Since this aspect of the experiment is so crucial and has often proven to be problematic, it is important to demonstrate the feasibility of the strategy prior to beginning a project. Genotyping in ES cells often requires a genomic Southern blot analysis that demonstrates the utility of the probe prior to the initiation of any ES cell work. Aspects to consider when designing a strategy for Southern blot analysis include:

- Diagnostic restriction enzymes should be chosen to give obvious size differences between targeted and non-targeted cell clones.
- 5′ and 3′ external probes for Southern analysis should give good signals on genomic Southerns. Both probes should be outside of the regions of homology used in the targeting vector (Fig. 2).

Alternatively, one can use PCR primers where one primer is outside of the region of homology, and the other is in a unique region of the targeting construct such as within the selectable marker (the gene encoding neomycin resistance, or hygromycin resistance, or puromycin resistance). This strategy is highly dependent upon long-range PCR, and thus, may be prone to error. To verify that this PCR analysis will work, one should generate a longer construct to test the PCR reaction that includes the region outside of the homologous arm of the targeting construct from which the external PCR primer was chosen [6] (Fig. 2). This construct can then be mixed in a single copy amount with genomic ES cell DNA for PCR verification prior to performing the targeting experiment. An advantage of using the PCR reaction is that it is easily translatable to the mouse tissues after mice are produced.

Once correctly targeted clones are identified, they are grown and then these ES cells are injected into mouse blastocysts. Most targeting facilities use blastocysts with genes for coat color markers that differ significantly from the coat color genes present in the ES cells [7]. Thus, when pups are born, they will be a mixture of cells from the blastocyst (with coat color X) and of ES cells (with coat color Y, dominant to coat color X). The investigator should backcross these chimeras to mice with coat color X. Any pups born with coat color Y will have to have been derived from ES cells. This is an

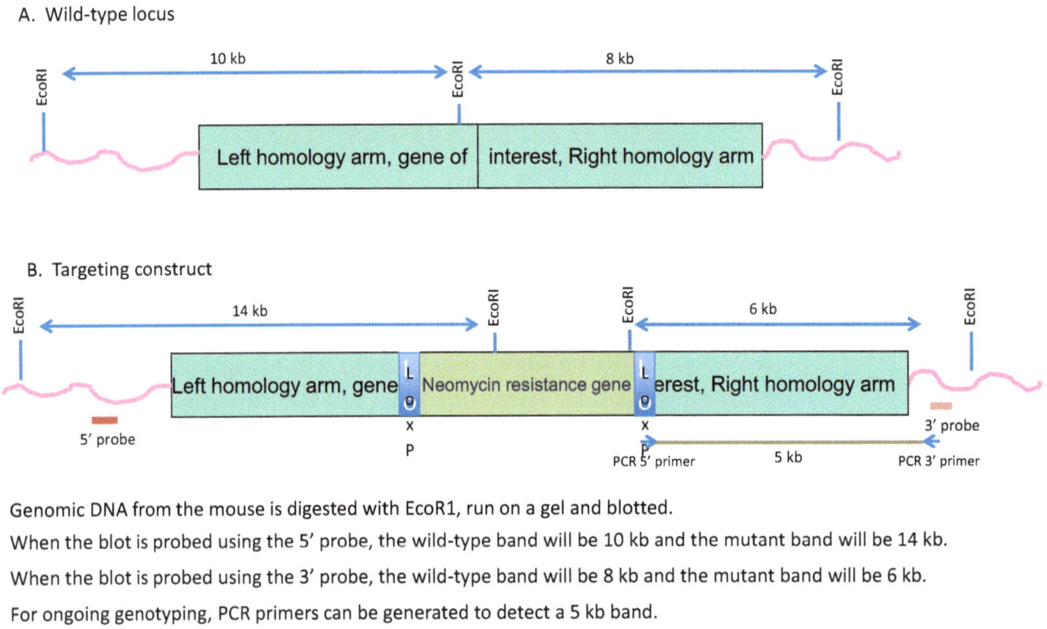

Fig. 2 Genotyping strategy to identify and distinguish a targeted allele from the endogenous wild type allele in ES cells and/or mice. (a) Wild-type allele of a gene of interest with EcoRI sites of 10 kb (left homology site) and 8 kb (right homology site). (b) When the gene for antibiotic resistance (in this case, Neomycin resistance) is inserted, additional EcoRI sites are added in the targeted allele, and probes can be designed outside of the homology arms that will give unique restriction fragments of 14 kb (left homology arm) and 6 kb (right homology arm). These fragments are easily distinguished from the wild-type restriction fragments. Additionally, PCR primers can be designed that will amplify a unique fragment that spans from the antibiotic resistance gene to an area outside of the homology region (brown fragment of 5 kb)

easy way to test for germ line transmission of the ES cell genotype. However, a caveat is that ES cells are diploid, and generally heterozygous for the targeted mutation, whereas the components of the germ line being tested are haploid. Thus, pups born with coat color Y will still need to be genotyped to determine which allele (wild-type or mutant) was transmitted. This can be done by genomic Southern blot or by PCR analysis.

Ongoing genotyping of the mice can evolve to the simpler PCR procedure. However, there are caveats to consider when one is genotyping the mice:

- Do not genotype solely for the selectable marker. During gene targeting, there can be random insertion of the construct in addition to the homologous insertion. These are generally not identified, and so can be easily mistaken for the correct homologous insertion if the genotyping is for the selectable marker, or the PCR primers reside only within the targeting construct.
- One should devise gene-specific primers and/or PCR across a relevant mutation (i.e., the point mutation made), and then sequence to ensure that it is correct.

- It is feasible to use real-time PCR [8] or PCR and melting techniques [9] to ensure that a mutation is present, but this technology should be verified by other means during the first couple of tests.

1.3 Gene-Targeted Mice Using CRISPR/Cas9 Technology

Genotyping of targeted animals produced using GEN strategies (i.e., CRISPR/Cas9) can be more problematic. With CRISPR/Cas9, the targeting takes place directly in the embryo by injection of a guide RNA and Cas9 enzyme (and a repair template if performing a knock-in) [1]. The guide RNA forms a complex with the Cas9 enzyme that then scans the DNA for sequences homologous to the 20 base unique sequence present in 5' of every guide RNA followed by a PAM (Protospacer adjacent Motif) sequence (NGG for spCas9). When that sequence is found, the guide stops and the Cas9 enzyme makes a double-stranded cut in the DNA [10]. CRISPR/Cas9 is used only to find and cut the DNA, so it leaves no footprint that can be used for genotyping. Once a double-stranded cut is made in the DNA, the cell's own repair mechanisms ligates the broken strands by immediately repairing the cut (leaving no mutation), or by trimming the DNA to make ends that can then be fused to repair the cut (leading to a deletion), or inserting nucleotides to allow for repair (leading to an insertion), or by inserting a DNA supplied by the investigator to repair the DNA (homology driven repair) [11] (Fig. 3).

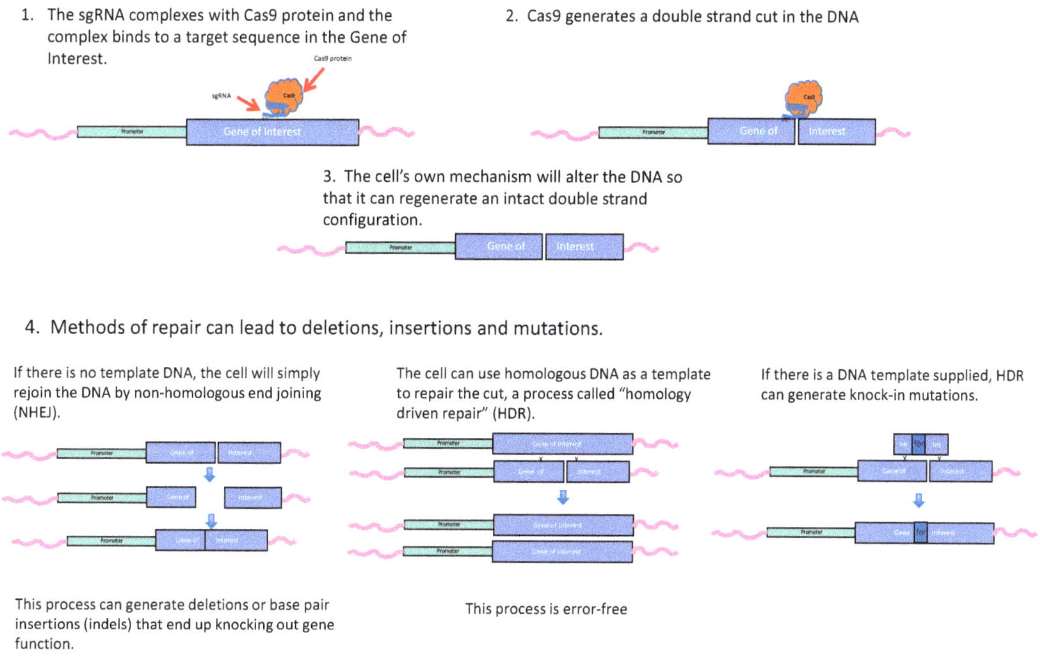

Fig. 3 CRISPR/Cas9 Gene Targeting

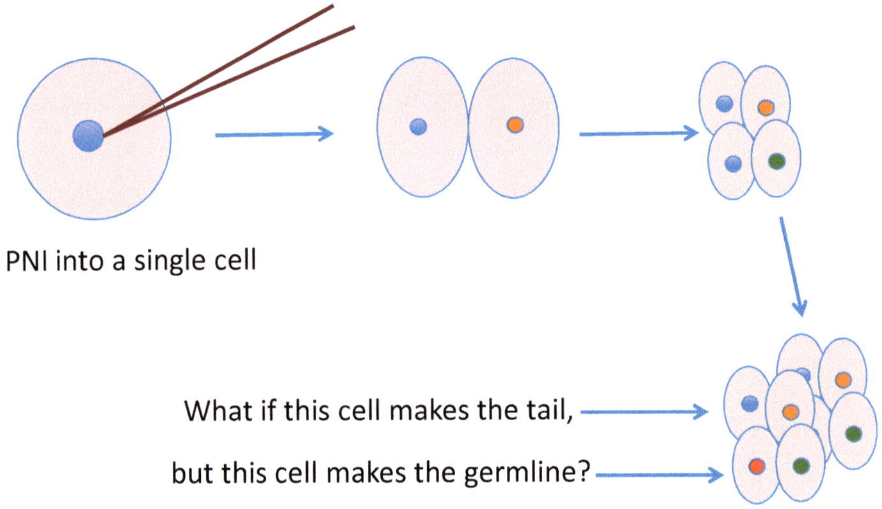

Fig. 4 How mosaics are generated by Cas9. Since Cas9 can act over several cell generations, there can be some cells in the G0 animal that do not have a mutation (blue nucleus), some that have an insertion (orange nucleus), and some that have a deletion (green nucleus). Even if no mutation is seen upon genotyping the G0 animals, it is possible to have a mutation transmitted through the germline to be detected in G1 pups

Depending on the cell division in the embryo when the mutation was induced, one may see cells with different mutations in the animal. Such an animal is known as a mosaic (cells from the same organism with different mutations in the DNA). The mosaicism observed in CRISPR/Cas9-edited animals likely happens due to the period of time over which the Cas9 nuclease acts [2, 12]. It may cut the DNA (one of the two alleles) in the first cell division leading to a deletion in one of the two daughter cells followed by cutting the DNA in the next cell division leading to a deletion or insertion (but perhaps only in one cell). You can see that, over several cell divisions, different cells may arise with differing mutations. The initial mice produced (G0 mice) could transmit any or all of these mutations in the germ line (Fig. 4).

Genotyping an animal where different cells may have different mutations can be challenging, to say the least. Initially, one can obtain DNA from the G0 mice, perform a PCR reaction across the region of interest, look for obvious changes such as large deletions or insertions, and then use the DNA generated from the PCR for sequencing to determine the changes that have occurred. If multiple mutations have happened, one can expect to see a standard sequence trace interrupted by a trace with multiple changes that then resolves once again to a single trace (Fig. 5). If this is seen, then it is essential to mate these mice to get a single mutation transmitted through the germ line, much like is done with ES cell generated chimeras. Thus, definitive genotyping of CRISPR/Cas9-generated mice should take place in the second (G1) generation, when the mutation has demonstrated germ line transmission.

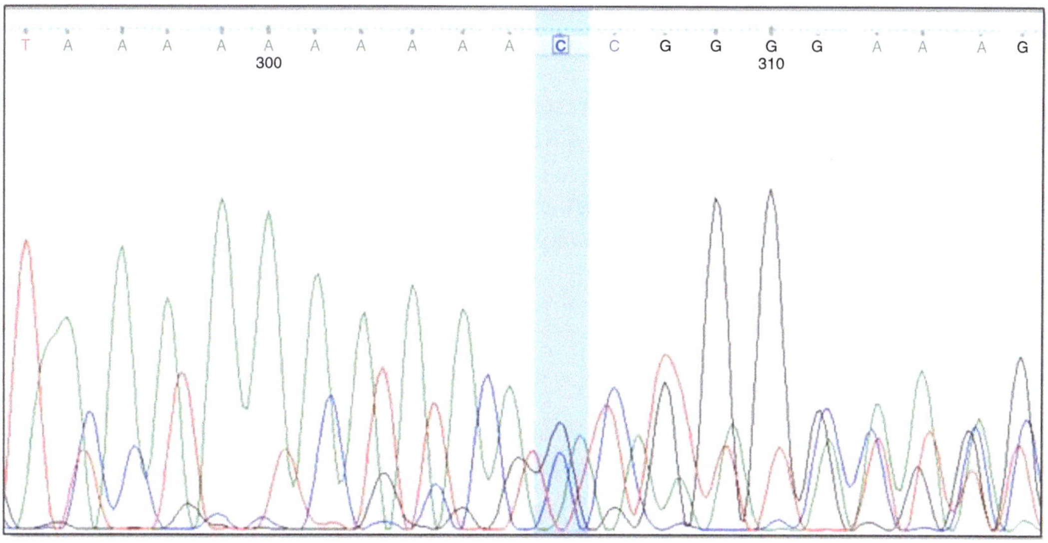

Fig. 5 Sequencing results from a **CRISPR/Cas9** G0 mouse where the donor cassette was designed to correct a point mutation at position 306. Note that there is both a G and a C at 306. As well, note that at position 314, all four nucleotides could be found

Regardless of the type of gene modification, once animals are produced and the lines established, there is ongoing need for genotyping. Transgenic mice are often maintained as heterozygotes, so pups will need to be genotyped at each generation. Targeted mice may exhibit embryonic lethality as homozygotes, and so need to be maintained as heterozygotes that will need to be genotyped at each generation. With good fortune, homozygous lines can be established for each type of model. However, even then, occasional genotyping is in order to ensure that no genetic contamination or loss of the mutation has occurred.

Whether a line is being used for the acquisition of preliminary data for a grant, has already been published, or is now being shared with other scientists, it is essential that the genotype expected is that being transmitted. In the USA, there is a new NIH requirement that an investigator should define how s/he will authenticate biological materials, and this must be detailed when submitting requests for grant funding [13]. With regard to animal models produced, this would logically extend to verification of the mouse model by genotyping. In addition, if one receives an animal model from another investigator, it is in his/her best interest to verify the genotype prior to using the mouse in an experiment. Many an experiment has produced unexpected results or no results and been discarded when, in fact, it was the genotype of the animal being tested that was incorrect.

2 Materials

2.1 Simple PCR Analysis of Small Amounts of Tissue

A simple PCR reaction can be used to genotype pups, genotype blastocysts after CRISPR/Cas9 targeting to test sgRNA cutting, or to genotype ES cells. This protocol is NOT FOR USE when large amounts of DNA are required (e.g., genomic Southerns).

Prepare all solutions using PCR grade water. Buffer solutions can be maintained at room temperature. The PCR reaction mix should be maintained on ice.

1. 10× PCR Buffer: 500 mM KCl, 100 mM Tris–Cl pH 8.3, 25 mM $MgCl_2$.
2. Tissue Digestion Buffer: 10× PCR buffer diluted to 1× with dH_2O, 0.45% IGEPAL, 0.45% TWEEN20.
3. 10 mg/mL Proteinase K.
4. PCR Reaction Mix (20 μL/sample): 200 μM dNTPs, 1 μM each primer, 1 U Taq polymerase/each 20 μL reaction, 1× PCR buffer (WITHOUT detergents) to obtain volume of 20 μL/sample.

2.2 DNA Preparation for Southern Blot and CRISPR/Cas9 Screening

This protocol results in cleaner DNA and can be used for PCR for genotyping as well. Prepare all solutions using sterile water. Buffer solutions can be maintained at room temperature.

1. Lysis Buffer: 10 mM Tris pH 8.0, 100 mM NaCl, 10 mM EDTA pH 8.0, 0.5% SDS.
2. 3 M Sodium Acetate (NaOAc): 408.24 g/L in H_2O, pH 5.2.
3. Proteinase K: 20 mg/mL in H_2O.

2.3 Screening CRISPR Mice

Many methods have been used to screen CRISPR/Cas9-targeted mice/cells. If a new restriction site is introduced or an existing site is disrupted, a simple restriction digest can be used for screening. If a new DNA sequence (loxP, GFP, etc.) is introduced, PCR can be used for size-based screening. For gene knock-out and point mutation knock-in projects, the Surveyor Assay can be used [14].

For a manageable number of samples (<20), the best approach is to PCR amplify a 1 kb region (0.5 kb on either side of the targeted site) and sequence it. Off-site targeting can be screened using the Surveyor Assay and/or sequencing of top ten scoring off-target sites. Alternatively, whole genome sequencing can be done to look at all possible off-target sites.

1. PCR Reaction Mix (20 μL/sample): 50–200 ng DNA, 200 μM dNTPs, 1 μM each primer, 1 U Taq polymerase/each 20 μL reaction, 1× PCR buffer (WITHOUT detergents) to obtain volume of 20 μL/sample.

2. Restriction Digest Reaction: 500 ng to 1 μg DNA, 3 μL of 10× Buffer, 10 U of desired restriction enzyme, sterile H_2O up to 30 μL final volume, 1% agarose gel, TAE buffer.
3. Surveyor Assay: 500 ng to 1 μg DNA, 3 μL of 10× Buffer, 10 U of Surveyor nuclease, sterile H_2O up to 30 μL final volume, 1% agarose gel, TAE buffer.
4. Sanger Sequencing: 100 ng/μL DNA, 1 μM forward or reverse primer.

2.4 Restriction Digest and Gel Electrophoresis for Southern Blot

Southern blotting is rapidly falling out of favor due to the complexity of the procedure and the fact that large amounts of radioactive liquids are generated. However, this is the gold standard procedure for initial assay of targeted embryonic stem cells.

1. Restriction Digest Reaction: 10 μg of clean genomic DNA, 10 μL of 10× Buffer, 30 U of desired restriction enzyme, sterile H_2O up to 100 μL final volume.
2. Gel Separation of Digest: 1% agarose gel, 1× TAE.

2.5 Southern Blot Transfer

1. Depurination Solution (0.25 M HCl): 10.77 mL conc. HCl, sterile H_2O up to 500 mL.
2. 5 M NaCl solution: 292.2 g NaCl, sterile H_2O up to 1 L.
3. Denaturation Solution: 10 g NaOH, 150 mL of 5 M NaCl solution, sterile water up to 500 mL.
4. Neutralization Solution: 150 mL 5 M NaCl, 125 mL 2 M Tris, pH 7.6, sterile H_2O up to 500 mL.
5. 20× SSC Solution: 175.3 g of NaCl, 88.2 g of Sodium Citrate, sterile H_2O up to 1 L, pH 7.0.
6. 20× SSPE Solution: 175.3 g NaCl, 27.6 g $NaH_2PO_4{\cdot}H_2O$, 7.4 g $Na_2EDTA{\cdot}2H_2O$, sterile H_2O up to 1 L, pH 7.4.
7. Pre-Hybridization Buffer: 30 mL 20× SSPE, 20 mL 50× Denhardt's Solution, 1% (w/v) SDS, sterile H_2O up to 100 mL.
8. Hybridization Buffer: 30 mL 20× SSPE, 1% (w/v) SDS, sterile H_2O up to 100 mL.
9. Wash Buffer 1: Dilute 20× SSPE to 2×, 0.1% SDS.
10. Wash Buffer 2: Dilute 20× SSPE to 0.1×, 0.1% SDS.

3 Methods

3.1 Simple PCR Analysis of Small Amounts of Tissue

1. Obtain tissue sample and store frozen until use. This protocol is designed for use with very small amounts of tissue (1–2 mm tail biopsy, 2 mm ear clip, 1 mm toe clip, blastocyst).

2. To each tissue sample, add 100 μL of the Tissue Digestion Buffer (5 μL for single blastocysts) and 10 μL of Proteinase K (10 mg/mL) (0.5 μL for single blastocysts).
3. Place the sample at 60 °C for 2 h to overnight. For single blastocysts, you may wish to overlay the sample with mineral oil to reduce evaporation. If using a short incubation time, vortex the tube several times during the incubation. If the procedure needs to be suspended, this is an excellent time to do so.
4. After tissue digestion and just prior to the PCR reaction, denature the Proteinase K in the solution by boiling for 10–15 min, then do a quick spin down to collect the liquids in the bottom of the tube. Immediately place sample on ice (*see* **Note 1**).
5. Aliquot 18 μL of the PCR reaction buffer into a PCR tube.
6. Add 2 μL of the tissue digest to the tube and mix by tapping or pipetting.
7. Place into thermocycler and run; use 30 cycles with the annealing temperature set to optimize annealing using your specific primers.
8. When PCR is complete, add 2 μL of gel loading dye and run on a 1.5–2% agarose gel.

3.2 DNA Preparation for Southern Blot and CRISPR/Cas9 Screening

1. Cut a small piece (2–4 mm) from the tail tip of a 3-week-old mouse and place in a clean 1.5 mL Eppendorf tube. For ES cells, collect the cells and transfer to a 1.5 mL Eppendorf tube.
2. Add 500 μL of Lysis Buffer and 20 μL of Proteinase K to each tube.
3. Incubate samples in 55 °C water bath overnight. For ES cells, incubate at 37 °C.
4. Vortex each tube for 10 s, and spin at maximum speed (13,200 rpm/16,100 rcf/*g*-force) for 10 min at room temperature (RT).
5. Transfer supernatant to a new tube.
6. Add equal volume of PCIA Solution (phenol:chloroform:isoamyl alcohol:: 25:24:1), vortex for 10 s, and spin at max speed for 5 min.
7. [Optional step] If there is the possibility of phenol carryover, you can perform an optional chloroform-only extraction using 1:1 chloroform: sample. Add chloroform, vortex for 10 s, and spin at max speed for 5 min.
8. For both **steps 6** and 7, transfer top layer to a new tube. Place tube on ice.
9. Add 1/10th volume of 3 M NaOAc and 2× volume of cold 100% ethanol.

10. Incubate at −20 ° C for 30 min to overnight.
11. Spin at max speed for 30 min at 4 °C.
12. Discard supernatant, and wash pellet with 70% ethanol twice at 4 °C.
13. Air dry the pellet and resuspend in 200 μL sterile H_2O or TE by pipetting up and down several times. Incubate at RT for 1–2 h. Vortex for 10 s.
14. Check the DNA concentration using Nanodrop.

3.3 Screening CRISPR Mice

3.3.1 Restriction Digest Based Screening

1. PCR amplify 1 kb region at target site.
2. Aliquot 48 μL of the PCR reaction buffer into a PCR tube.
3. Add 2 μL of the tissue digest to the tube and mix by tapping or pipetting.
4. Place into thermocycler and run; use 30 cycles with the annealing temperature set to optimize annealing using your specific primers.
5. When PCR is complete, run 5 μL of each sample on a 1% agarose gel.
6. If a single expected band is present, purify the remaining PCR by using QIAquick PCR purification kit or similar product. If multiple bands are present, run the remaining PCR in a 1% agarose gel, cut out the expected band, and extract DNA using QIAquick Gel Extraction Kit or similar product.
7. Check DNA concentration using Nanodrop.
8. Digest 1 μg of PCR product with the restriction enzyme.
9. Aliquot required volume (to make 1 μg) of purified PCR DNA into a tube.
10. Add 3 μL 10× Buffer and 10 U of restriction enzyme.
11. Add sterile H_2O to make up 30 μL final Volume.
12. Incubate digest at required temperature for 1–2 h.
13. Run the sample on a 1–2% agarose gel (choose agarose% that will best separate your bands).
14. Positive samples should be sequenced (using the remaining purified PCR DNA) to identify the exact mutational changes.

3.3.2 Size-Based Screening for Insertion of Long DNA Sequences

1. PCR amplify DNA fragment of appropriate size. For example, insertion of 40 bp loxP sequence can be screened by amplifying 0.2–0.3 kb fragment. Insertion of large fragments can be screened by amplifying any sized DNA fragment.
2. Aliquot 48 μL of the PCR reaction buffer into a PCR tube.
3. Add 2 μL of the tissue digest to the tube and mix by tapping or pipetting.

4. Place into thermocycler and run; use 30 cycles with the annealing temperature set to optimize annealing using your specific primers.
5. When PCR is complete, run 5 μL of each sample on a 1% agarose gel.
6. If a single expected band is present, purify the remaining PCR by using QIAquick PCR purification kit or similar product. If multiple bands are present, run the remaining PCR in a 1% agarose gel, cut out the expected band, and extract DNA using QIAquick Gel Extraction Kit or similar product.
7. Run 500 ng to 1 μg PCR DNA on agarose gel of appropriate percentage.
8. Positive samples should be sequenced (using the remaining purified PCR DNA) to identify the exact mutational changes.

3.3.3 Surveyor Assay

1. PCR amplify 1 kb fragment from the test DNA sample as well as a DNA sample from a wild-type (WT) (non-mutated) mouse.
2. Aliquot 48 μL of the PCR reaction buffer into a PCR tube.
3. Add 2 μL of the tissue digest to the tube and mix by tapping or pipetting.
4. Place into thermocycler and run; use 30 cycles with the annealing temperature set to optimize annealing using your specific primers.
5. When PCR is complete, run 5 μL of each sample on a 1% agarose gel.
6. If a single expected band is present, purify the remaining PCR by using QIAquick PCR purification kit or similar product. If multiple bands are present, run the remaining PCR in a 1% agarose gel, cut out the expected band, and extract DNA using QIAquick Gel Extraction Kit or similar product.
7. Check DNA concentration using Nanodrop.
8. Mix 500 ng of sample PCR DNA with 500 ng of WT PCR DNA. Add sterile H_2O to make final volume of 25 μL.
9. Denature the combined DNA sample at 95 °C for 5 min. Incubate at room temperature for 2 h.
10. Add 3 μL of 10× buffer and 10 U of surveyor nuclease. Adjust to a 30 μL final volume with sterile H_2O. Surveyor nuclease only cleaves heterozygous DNA (WT and mutant strand hybridization) at the site of mismatch giving two smaller fragments.
11. Run samples on a 1% agarose gel.
12. Positive samples should be sequenced (using the remaining purified PCR DNA) to identify the exact mutational changes.

3.3.4 Sequencing

1. Design primers 0.5 kb upstream and downstream of target site.
2. PCR amplify the 1 kb region at target site as noted in the "Surveyor Assay."
3. When PCR is complete, run 5 μL of each sample on a 1% agarose gel.
4. If a single expected band is present, purify the remaining PCR by using QIAquick PCR purification kit or similar product. If multiple bands are present, run the remaining PCR in a 1% agarose gel, cut out the expected band, and extract DNA using QIAquick Gel Extraction Kit or similar product.
5. Check DNA concentration using Nanodrop.
6. Sequence the DNA using either primer.
7. If sequencing result is unclear (generally because of mosaicism), the PCR products can be subcloned into a plasmid, grown, and DNA isolated.
8. The plasmid subclones are then sequenced to identify all events.

3.4 Restriction Digest and Gel Electrophoresis for Southern Blot

1. Digest 10 μg of genomic DNA with desired restriction enzyme for 4 h to overnight (consider the length of time based on enzyme's star activity).
2. Add 20 μL of 6× DNA loading dye (commercially available).
3. Run the sample on a 1% agarose gel for 3 h at 80 V.
4. Place a ruler next to gel and take a picture of the gel with size markers.

3.5 Southern Blot Transfer

1. Depurination (Optional for fragments less than 10 kb, breaks DNA into smaller fragments for easier transfer): Transfer the gel to a tray and submerge in 0.25 M HCl solution. Gently shake/rotate for 8 min. Discard the solution and rinse the gel with water twice.
2. Denaturation (denatures DNA into single-strands): Add denaturation buffer and shake/rotate for 15 min. Repeat once and rinse with water.
3. Neutralization: Add neutralizing buffer and shake/rotate for 30 min. Repeat one time.
4. DNA transfer to membrane: Take a large tray, fill it with 10× SSC (dilute 20× SSC 1:1 with water) and place a glass plate across the top to form a bridge from one side to the other. Alternatively, a platform can be placed inside the tray (Fig. 6).
5. Make a wick using blotting paper (2 ft. × 8 in.) that goes over the glass plate/platform into the glass tray. Soak it with 10× SSC.
6. Place gel (well side down) on top of the paper wick and remove air bubbles.

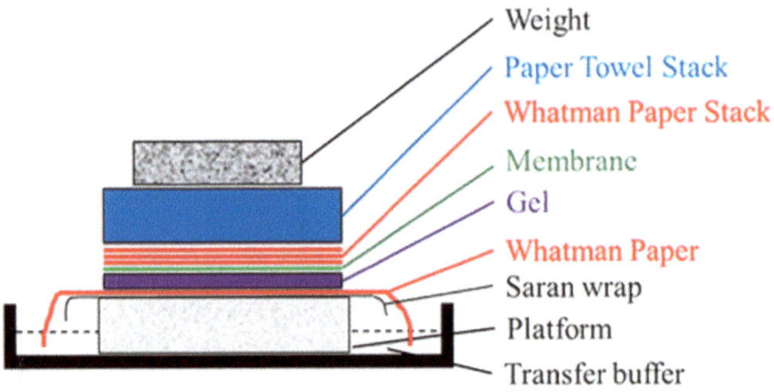

Fig. 6 Southern Blot setup (Source: http:/www.sigmaaldrich.com/technical-documents/articles/biology/southern-and-northern-blotting.html)

7. Cut a membrane and two blotting papers that are the same size as the gel, and soak them in 2× SSC (dilute 20× SSC 1:10 with water). Label on the upper right corner of the membrane with pencil.
8. Place the membrane on top of the gel and remove air bubbles using a roller.
9. Add the blotting papers on top of the membrane followed by stacks of paper towels. Do not allow the blotting papers or the paper towels to come in contact with the wick. Place a weight on top and cover the entire assembly with Saran wrap.
10. Leave overnight.
11. The following day, retrieve the membrane, and place it inside a UV-cross-linker on top of a blotting paper. Cross-link with automatic settings.

3.6 Radiolabeled DNA Probe Preparation

1. PCR amplify the desired DNA fragment to be used as a probe.
2. Run the sample on a 1% agarose gel, cut out the correctly sized band, and extract the DNA using QIAquick Gel Extraction Kit, Catalog no. 28706, or other available product.
3. Check the DNA concentration.
4. Follow protocol of the desired radiolabeling kit to obtain radiolabeled DNA probe. Label the probe DNA to a specific activity of $>1 \times 10^8$ dpm/µg [15] (check radioactivity; the signal should be very strong) (*see* **Note 2**).

3.7 Pre-hybridization and Hybridization

1. Pre-hybridization: Place the membrane into a hybridization tube or a sealable bag. Add 1 mL of pre-hybridization buffer with 50 µg/mL of denatured salmon sperm DNA for each 10 cm^2 of the membrane. Close the tube or seal the bag.

Incubate the membrane in the hybridization solution for 1–2 h at 65 °C. If the hybridization buffer contains formamide, incubate at 42 °C.

2. Denature radiolabeled DNA probe at 95 °C for 5 min. Transfer the tube into ice immediately.
3. Hybridization: Carefully open the hybridization tube or cut open a small corner of the sealed bag. Pour off the pre-hybridization buffer. Add pre-warmed hybridization buffer with radiolabeled probe (25 ng DNA probe/mL of pre-warmed hybridization buffer). The amount of probe and buffer needed will depend on the size of the membrane and the container used, typically 1 mL of hybridization buffer for each 10 cm^2 of the membrane [15]. If using a sealable bag, remove all bubbles from the bag and re-seal the bag.
4. Hybridize overnight at 65 °C (42 °C for buffer with formamide) in a hybridization oven, or, if using a sealed bag, a gently shaking water bath.
5. The following day, carefully discard the buffer in the designated container for radioactive waste.
6. Wash for 15 min with Wash Buffer 1 at room temperature. Rotate or gently shake the container with the membrane and wash buffer. Repeat once. Discard each wash into the radioactive waste.
7. Wash for 15 min with Wash Buffer 2 at 55–65 °C in the hybridization oven or shaking water bath. As the washes will remove more DNA with higher temperatures, you should consider the G/C content of your probe to determine whether to start at a lower or higher temperature ($\%\ A/T > \%\ C/G$ = lower temperatures for washing). Repeat once. Discard each wash into the radioactive waste.
8. Check the signal with Geiger counter. The top and bottom parts of the membrane should not have any signal. The signal should be localized to the area with the expected size of DNA fragment. If unwanted signal is present, repeat **step 8**.
9. Expose the membrane to X-ray film or a Phosphor screen.

4 Notes

1. If the sample is allowed to slowly come to room temperature, rather than being placed on ice, the Proteinase K can reform and digest the Taq polymerase and no band will be seen.
2. A high specific activity of the probe is essential for genomic Southerns to obtain a good signal.

Acknowledgments

The authors would like to thank Dr. Vinod Pant for sharing his expertise with Southern blot analysis. This work was supported by a Cancer Center Support Grant NCI # CA016672(GEMF).

References

1. Singh P, Schimenti JC, Bolcun-Filas E (2015) A mouse geneticist's practical guide to CRISPR applications. Genetics 199:1–15. https://doi.org/10.1534/genetics.114.169771
2. Yen ST, Zhang M, Deng JM, Usman SJ, Smith CN, Parker-Thornburg J, Swinton PG, Martin JF, Behringer RR (2014) Somatic mosaicism and allele complexity induced by CRISPR/Cas9 RNA injections in mouse zygotes. Dev Biol 393:3–9. https://doi.org/10.1016/j.ydbio.2014.06.017
3. Behringer R (ed) (2014) Manipulating the mouse embryo: a laboratory manual, 4th edn. Cold Spring Harbor Laboratory Press, New York
4. Molto E, Vicente-Garcia C, Montoliu L (2011) Designing transgenes for optimal expression. In: Pease S, Saunders TL (eds) Advanced protocols for animal transgenesis. Springer, New York
5. Nagy A, Gertsenstein M, Vintersten K, Behringer RR (2004) Detection and analysis of mouse genome alterations and specific sequences. In: Manipulating the mouse embryo, 3rd edn. Cold Spring Harbor Laboratory Press, New York
6. Saunders TL (2011) Gene targeting vector design for embryonic stem cell modifications. In: Pease S, Saunders TL (eds) Advanced protocols for animal Transgenesis. Springer, New York
7. Brennan K (2011) Colony management. In: Pease S, Saunders TL (eds) Advanced protocols for animal transgenesis. Springer, New York
8. Hnatyszyn HJ, Podack ER, Young AK, Seivright R, Spruill G, Kraus G (2001) The use of real-time PCR and fluorogenic probes for rapid and accurate genotyping of newborn mice. Mol Cell Probes 15:169–175. https://doi.org/10.1006/mcpr.2001.0355
9. Thomsen N, Ali RG, Ahmed JN, Arkell RM (2012) High resolution melt analysis (HRMA); a viable alternative to agarose gel electrophoresis for mouse genotyping. PLoS One 7:e45252. https://doi.org/10.1371/journal.pone.0045252
10. Jinek M, Chylinski K, Fonfara I, Hauer M, Doudna JA, Charpentier E (2012) A programmable dual-RNA-guided DNA endonuclease in adaptive bacterial immunity. Science 337:816–821. https://doi.org/10.1126/science.1225829
11. Harms DW, Quadros RM, Seruggia D, Ohtsuka M, Takahashi G, Montoliu L, Gurumurthy CB (2014) Mouse genome editing using the CRISPR/Cas system. Curr Protoc Hum Genet 83(15.7):11–27. https://doi.org/10.1002/0471142905.hg1507s83
12. Richardson CD, Ray GJ, DeWitt MA, Curie GL, Corn JE (2016) Enhancing homology-directed genome editing by catalytically active and inactive CRISPR-Cas9 using asymmetric donor DNA. Nat Biotechnol 34:339–344. https://doi.org/10.1038/nbt.3481
13. Lauer M (2016) Authentication of key biological and/or chemical resources in NIH grant applications. http://nexus.od.nih.gov/all/2016/01/29/authentication-of-key-biological-andor-chemical-resources-in-nih-grant-applications/
14. Yang B, Wen X, Kodali NS, Oleykowski CA, Miller CG, Kulinski J, Besack D, Yeung JA, Kowalski D, Yeung AT (2000) Purification, cloning, and characterization of the CEL I nuclease. Biochemistry 39:3533–3541
15. Brown T (2001) Hybridization analysis of DNA blots. Curr Protoc Immunol. Chapter 10: Unit 10.16B. https://doi.org/10.1002/0471142735.im1006bs06

Chapter 13

Nomenclature: Naming Your Gene-Modified Mouse

Alicia Valenzuela

Abstract

For many of us, if we are honest, nomenclature is a tedious, incomprehensible jargon that interferes with presenting and reading research data. While understanding the rules governing nomenclature involves a steep learning curve, the curve is short, and the basics, with a little effort, are grasped relatively quickly.

Like a language, nomenclature is a communication tool that provides a common ground for a disparate group of people. Standardized names provide universally recognized identifiers that can be used by technicians, researchers, purchasing agents, and facility managers, in fact, anyone who uses mice. The formal nomenclature conveys information on the genetics, the technology involved in making the mutation, who created and maintained the strain, and its relationship to other strains. Using a standardized nomenclature for genes, alleles, and strains assists in the goal of reproducible science and helps to bridge the vast amount of data generated by multi-species genome projects.

Key words Mouse, Nomenclature, Standardized, Names, Mouse genetics, Mouse strain, Gene-modified, Reproducible science

1 Introduction

Dunn, Gruneberg and Snell published the first report of the Committee on Mouse Genetics Nomenclature in 1940. The Committee outlined a set of rules governing the naming of mouse strains and genes [1]. Their report reflected an understanding by early mouse geneticists that a common language was needed to describe the mouse models used in research. Mouse strain nomenclature provides in an astonishingly few number of letters (and numbers), a wealth of information about the genetic background, the mutation, and the origin of mouse strains. Understanding the basics of nomenclature requires a relatively minimal amount of effort, but it is effort well spent for both the individual and the scientific community. Using formal nomenclature supports reproducible science and ensures that investigators appropriately select and understand the mouse strains used in their research.

At its most basic, a mutant strain name consists of three parts: genetic background, genetic element of interest, and substrain.

Melissa A. Larson (ed.), *Transgenic Mouse: Methods and Protocols*, Methods in Molecular Biology, vol. 2066,
https://doi.org/10.1007/978-1-4939-9837-1_13,

These three parts are almost always separated by two punctuation marks, a hyphen (-) and slash (/), respectively. For example, the strain name: B6.129P2-*Apoe*^{tm1Unc}/J refers to the genetic background (B6.129P2): a C57BL/6 congenic background; the genetic element (*Apoe*^{tm1Unc}): a targeted mutation in the gene, *Apoe*; and the substrain/labcode (J): The Jackson Laboratory. This chapter will describe each of these elements, giving the user the tools to name their own gene-modified mice.

Nomenclature guidelines are established by The International Committee on Standardized Genetic Nomenclature for Mice. The committee meets once a year to review and revise mouse nomenclature. Their decisions are incorporated into The Mouse Genome Database (MGD), the authoritative source of official names for mouse genes, alleles, and strains. The full set of rules and guidelines is available online [2].

2 Inbred Strains, Laboratory Codes, Hybrids, and Stocks

2.1 Inbred Strains

Any discussion of mouse nomenclature begins with inbred strains. First developed in 1909, inbred strains represent the keystone of mouse genetics. Inbred strains are defined as the product of sister x brother ("filial") mating for 20 or more consecutive generations (F = 20). An individual mouse from an inbred strain can be traced back to a single ancestral pair. At F20, approximately 98.6% of the paired alleles in each mouse are homozygous [3]. Most commonly used inbred strains have been inbred for over 200 generations, making them essentially genetically identical or "isogenic." It is this uniformity that makes inbred strains highly valuable in scientific research. However, the process of inbreeding is not perfect. Residual heterozygosity, genetic contamination, copy number variation, and genetic drift can reduce homozygosity to less than 100%.

Each inbred strain carries a unique set of traits that are considered to be characteristics of the strain. Some strains, through common use and extensive genetic characterization, have become all-purpose research mice. Other inbred strains were selectively bred for certain characteristics, including susceptibility to cancer, diet-induced obesity, hypertension, and numerous other complex traits. There are approximately 400 inbred strains and most of these were created in the early to mid-twentieth century [4]. A listing of available inbred strains can be found at the website of the International Mouse Strain Resource (IMSR) [5]. Few inbred lines are created today due to cost and time constraints. New strains that are developed are often bred selectively for two contrasting polygenic traits such as long sleep/short sleep (ILS/ISS), alcohol withdrawal convulsion severity (IWSP1/IWSR1), and blood pressure high/low (BPH/BPL).

Detailed information about inbred strains is available through individual Repository webpages and Festing's Inbred Strains of Mice and their Characteristics (an older, but helpful guide) on the Mouse Genome Informatics website [4]. The Mouse Phenome Database provides a useful resource both for single nucleotide polymorphism (SNP) data and for standardized phenotypic data [6, 7].

An inbred strain is designated first by the parent (or root) strain, a unique, brief symbol consisting of upper case roman letters or a combination of letters and numbers beginning with a letter, such as C57BL, C3H, and DBA, and followed by a slash (/) and the substrain designation. There are always exceptions to any rule, and, in this case, there are some well-known strains that do not follow this convention and begin with numbers (e.g., 129P3).

Although inbred strain names appear to be a random mix of letters and numbers, there is often historical context for the name. Names can be based on coat color (C57BL/6—black), origin (NZB—New Zealand Black) or phenotype (NOD—nonobese diabetic). The second part of an inbred strain name is the substrain and/or laboratory code designation.

2.2 Substrains

Genetic drift refers to the random fluctuation in allele frequencies over time. Spontaneous mutations, mostly neutral and undetected, randomly appear in inbred populations and can become fixed. When colonies of an inbred strain are maintained separately, mutations or allelic variants can cause genetic divergence between colonies (or branches) of an inbred strain. A branch that is either known or suspected to be genetically different from the original inbred strain is designated a "substrain." The criteria for designating a substrain are: mice from an inbred strain are maintained separately from the original or foundation strain for more than 20 additive generations; fixed genetic differences are detected between the original strain and the separated line; mice are separated from the original colony after F20, but before F40.

In the nomenclature, a substrain is designated by the addition of a slash (/) to the root (or parent) name followed by the laboratory code of the holder (discussed in the next section). In some instances, notably C57BL, a line number is included (Fig. 1).

2.3 Laboratory Codes

A Laboratory Registration Code or lab code is a series of (1–5) letters, the first capitalized, which uniquely identifies a particular institution or investigator's laboratory. For example, the lab code for Dr. Bruce Beutler is Btlr. Lab codes are registered with the Institute for Laboratory Animal Research (ILAR), which maintains a master registry [8]. Lab codes are used to identify the originators of mutations and transgenes and the producers and subsequent holders of inbred, congenic, and other strains of laboratory animals. Lab codes serve to differentiate between independently

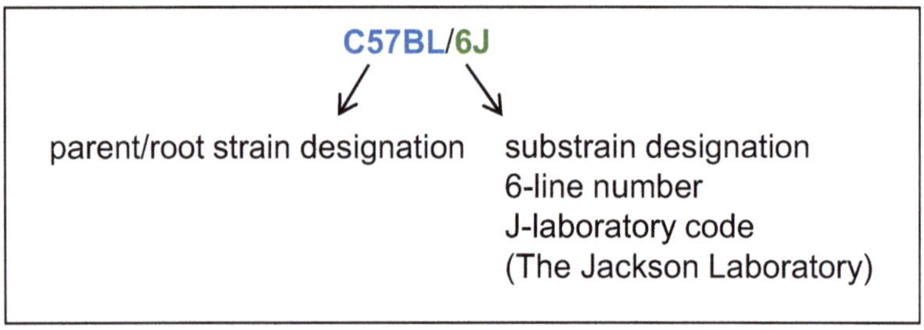

Fig. 1 Diagram of inbred strain nomenclature

Distinguishing between substrains:
C57BL/6J – The Jackson Laboratory
C57BL/6N – NIH
C57BL/6Ha – Hauschka (Roswell Park)
C57BL/6By– Bailey

Identifying substrain lineages:
129S4/SvJaeSor – Leroy Stevens (Sv) to Rudolf Jaenisch (Jae) to Philippe Soriano (Sor)

Identifying creators of engineered mutations:
$Cd44^{tm1Ugu}$ – first targeted mutation in *Cd44* from Ursula Gunthert (Ugu)
$Dicer2^{em2Psv}$ – second endonuclease-mediated mutation in *Dicer2* from Petr Svoboda (Psv)
Tg(ANPEP)861Mmul– transgenic insertion 861 (founder line) from Mathias Muller (Mmul)

Fig. 2 Uses of laboratory codes

developed congenic strains whose names are otherwise identical. Lab codes concatenated at the end of the strain name indicate the succession of colony holders who maintained the strain for multiple generations, allowing users to track potential genetic drift. For example, C3H/HeJArc is a substrain of C3H maintained by Heston (He), then transferred to The Jackson Laboratory (J) and then to Animal Resources Centre (Arc). Figure 2 provides some examples of the uses of lab codes (Fig. 2).

2.4 F1 Hybrids

An F1 hybrid is the first filial (F) generation of offspring produced by mating animals from two different inbred strains. All F1s from the same cross are genetically identical. The F1 hybrid is designated by the upper case abbreviations of the parent strain names followed by "F1." Abbreviations for inbred strain names are listed in Table 1. Please note that this list is not comprehensive, and in some instances, substrain information is included in the abbreviation (B6J, B6N). Figure 3 provides an example of an F1 hybrid – the *maternal* strain is always listed first (Fig. 3). F1 hybrids are often used in breeding to provide hybrid vigor (i.e., increased disease resistance, better survival under stress, greater natural longevity, and larger litters).

Table 1
Abbreviations for common inbred strain names

Strain	Abbreviation[a]
129 strains (include subtype)	e.g., 129S1, 129X1
A strains	A
AKR strains	AK
B	C57BL
C57BL/6 strains	B6
C57BL/10 strains	B10
C57BR/CD	BR
BALB/c strains	C
C3H strains	C3
C57L	L
CBA	CBA
DBA/1 strains	D1
DBA/2	D2
HRS/J	HR
RIIS/J	R3
SJL	SJL or J

[a]If substrains have known genetic or phenotypic differences, the substrain is included in the abbreviation (BALB/cBy becomes CBy)

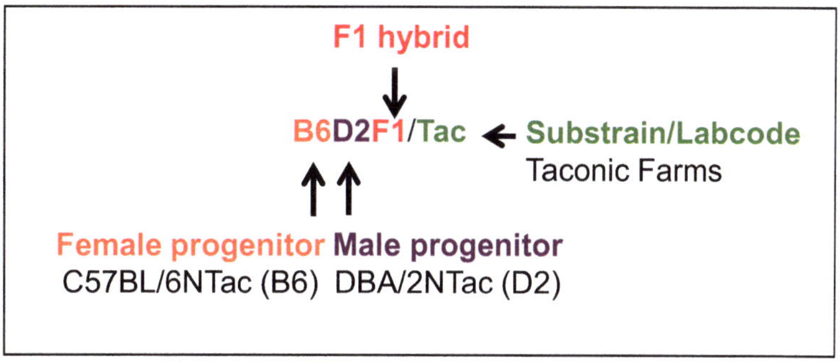

Fig. 3 Diagram of an F1 hybrid

2.5 Outbred Stocks

An outbred stock is a closed population (for at least 4 generations) of genetically variable animals that is bred to maintain maximum heterozygosity [9]. Outbreds are always referred to as "stocks" rather than strains. Outbred (also called random-bred) mice provide the advantage of approximating a natural heterogenous population by representing genetic diversity. They also tend to be healthier, have large litters and are less expensive. However, because they are segregating a variety of different alleles, they are not genetically identical and research using them may not be reproducible.

In outbred nomenclature, the lab code comes first, followed by a colon (:) and a stock name in capital letters. For example, Crl:SW is a Swiss Webster (SW) stock held by Charles River Laboratories (Crl).

3 Genetic Background in Mutant Strains

There are three basic parts of a mutant strain name: genetic background, genetic element of interest, and substrain/lab code (Fig. 4). These three parts are almost always separated by punctuation marks, a hyphen, and slash, respectively. This section describes the correct nomenclature to be used for different types of genetic backgrounds.

3.1 Coisogenic Strains

A coisogenic strain is an inbred strain in which a mutation was generated or arose spontaneously. A coisogenic strain is *never* crossed to another genetic background. In the following scenario, a transgenic construct (Tg(S100a4-cre)1Egn) is microinjected into fertilized BALB/c oocytes. Founder mouse 1 is crossed to BALB/c mice to establish the colony, and the colony is maintained as either a cross to BALB/c or as sibling mating. The formal strain name is BALB/c-Tg(S100a4-cre)1Egn/Yunk. This strain is on a "pure" genetic background, containing only one inbred strain. Coisogenic strains are ideal models for comparing the effects of mutations without confounding background effects. Unfortunately, it is not usually possible to compare multiple mutations on the same coisogenic backgrounds. The next best solution to control background effects is congenic strains.

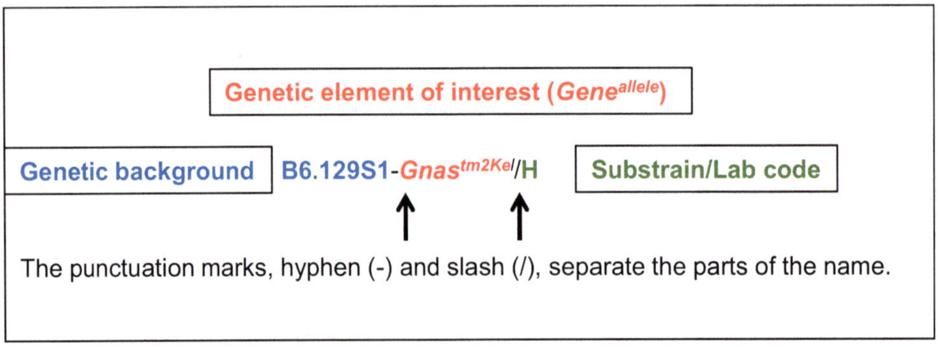

Fig. 4 Diagram of a mutant strain name

3.2 Congenic Strains

Congenic strains are produced by repeated backcrosses to an inbred strain (known as the "recipient," host, or background strain), with selection for a particular marker/allele from the original or "donor" strain (also known as strain of origin). The congenic strain and the inbred partner are expected to be almost identical at all loci except for the transferred locus and a linked segment of chromosome. The mutation is said to be introgressed into the recipient strain. An "incipient congenic" strain has been backcrossed between 5 and 9 times (N5-N9). A full congenic is backcrossed 10 or more times (N10). Statistically, at N5, 96.9% of alleles are from the recipient strain; at N10 this increases to 99.91% [10]. Alternatively, a congenic strain is produced using a marker-assisted "speed congenic" strategy. Offspring from each backcross are genotyped at a number of loci throughout the genome and only those with the fewest nonhost-derived alleles (or the most host-derived alleles) are backcrossed to produce the next generation. The speed congenic strategy can reduce breeding time from 10 generations to 4 or 5 generations. Figure 5 outlines the process of creating a congenic strain (Fig. 5). In the example, at least one cross should involve an inbred C57BL/6NTac male to replace the Y Chromosome. Notice that the punctuation in the genetic background changes from a semicolon (;) to a period (.) at N5 to indicate the transition from a mixed genetic background to an "incipient congenic" background. A period (.) in the strain name indicates a congenic strain. Full congenicity is anticipated after

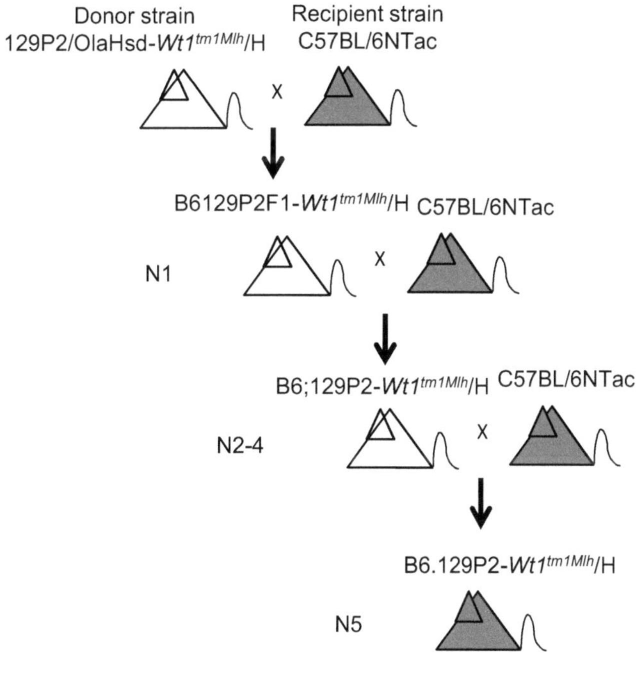

Fig. 5 Creation of a congenic strain

10 backcrosses to the inbred (N10 or greater). At the time of this writing, GigaMUGA (Giga Mouse Universal Genotyping Array), using more than 140,000 SNPs (single nucleotide polymorphisms), is increasingly used to more accurately assess the strain background.

The following are examples that illustrate more complicated scenarios.

In some cases, the "donor" strain (strain of origin) is an F1 or F2 hybrid. In this situation, it is usually not known which strain contributed to the region flanking and containing the gene of interest. B6.Cg-*Pvalb*$^{tm1.1(cre)Aibs}$ is an example of a targeted mutation (*Pvalb*$^{tm1.1(cre)Aibs}$) that originated in a (129S6/SvEv-Tac × C57BL/6NCrl)F1 hybrid (via the G4 ES cell line). Cg is used in the "donor" position to refer to "complex genome."

Names like B6.129S1(FVB)-*Gene*^{allele} reflect the following scenario: 129S1/Sv is the donor background (derived from an embryonic stem (ES) cell), mice from the colony were subsequently crossed to mice on an FVB/N background, and then to C57BL/6 for at least 5 backcrosses. The nomenclature indicates that some genetic background from the FVB/N inbred strain is present even though the strain is predominantly C57BL/6 in a nature. This situation often occurs when a targeted mutation is made using a construct that includes an FRT-flanked neomycin (neo) cassette. Often the neo cassette is removed by crossing to a strain carrying a flp recombinase, and this strain also contributes to the genetic background.

If, instead of FVB, an outbred strain was used or if, in addition to FVB, there was a cross to a fourth inbred, followed by the N5+ backcross to C57BL/6 then the name is B6.129S1(Cg)-*Gene*^{allele} and again, the Cg refers to "complex genome."

3.3 Mixed (Segregating) Strain Backgrounds

A strain whose genome is comprised of alleles derived from two inbred backgrounds, in which the alleles from each inbred represent more than 3% of the genome, is called segregating. If, in the example of the congenic strain in Fig. 5, the 129P2 donor is crossed to C57BL/6 less than 5 times (N2-4) then the name is B6;129P2-*Wt1*^{tm1Mlb}/H (Fig. 5). A *semicolon* between the two inbred strain name abbreviations indicates that the genetic background is a mixture of the two strains. The percent contribution of each strain is not conveyed in the nomenclature; however, the first strain listed represents the host strain in a donor/host relationship or represents a higher percentage of the strain background. The individual mice in this colony are not genetically identical; each carries a unique mix of the two contributing genomes.

One caveat is that if the strain has been maintained by sibling mating for greater than 20 generations (F20) then the strain is considered a recombinant inbred. In this case, the alleles representing the two inbreds are no longer randomly segregating but have become fixed. It is important to be aware of this situation although it will not be discussed in detail here.

3.4 Stocks

A mouse whose genome (excluding a congenic segment) is derived from *more than two* inbred strains or includes genetic contribution from an *unknown or outbred* source is considered a stock (not a strain). In this case, no attempt is made to incorporate the progenitor strain names; it is referred to as STOCK and the hyphen between genetic background and mutation is replaced by a space, as in: STOCK $Gnai1^{tm1Rne}$/N.

4 Gene Modifications: Mutations and Nomenclature

As discussed earlier, there are three basic parts in a mutant strain name: genetic background, genetic element of interest, and substrain/lab code (Fig. 4). The term, "genetic element of interest", covers a broad range of unique genome features including, but not limited to, cytogenetic markers, complex/cluster regions, and quantitative trait loci (QTL), but most commonly, refers to a transgene or a mutant or variant allele. Mouse gene symbols are italicized and the first letter capitalized. For example, the gene symbol for the interleukin 2 receptor gamma chain is *Il2rg*. Human gene symbols are italicized and use all capital letters, that is, *IL2RG*. Alleles are designated by italics and are superscripted to the gene symbol—$Gene^{allele}$. $Il2rg^{tm1Wjl}$ indicates that "*tm1Wjl*" (targeted mutation 1, Warren J Leonard) is an allele of the gene *Il2rg*.

4.1 Targeted Mutations

Targeted mutant strains carry deletions, additions, duplications, or other alterations to wild-type sequences that are introduced via homologous recombination by targeting a DNA construct into embryonic stem (ES) cells. Correctly targeted ES cells (often derived from 129 substrains) are then microinjected into or re-aggregated with host embryos. Germline transmission is confirmed following a cross between chimeric males (assuming the use of a male ES cell line) and inbred females (frequently C57BL/6). If the newly established line has a disrupted or deleted gene that prevents protein expression, it is commonly called a knockout (KO). If it has a new or duplicated genetic element that is expressed, it is called a knockin (KI).

In the nomenclature, a targeted mutation is designated by the superscript tm#Labcode, where "tm" indicates "targeted mutation," # is a serial number referring to the number of alleles of a given gene from a given laboratory, and the lab code (discussed previously) designates where the mutation was produced. A targeted mutation is an allele of the target gene; therefore, it is

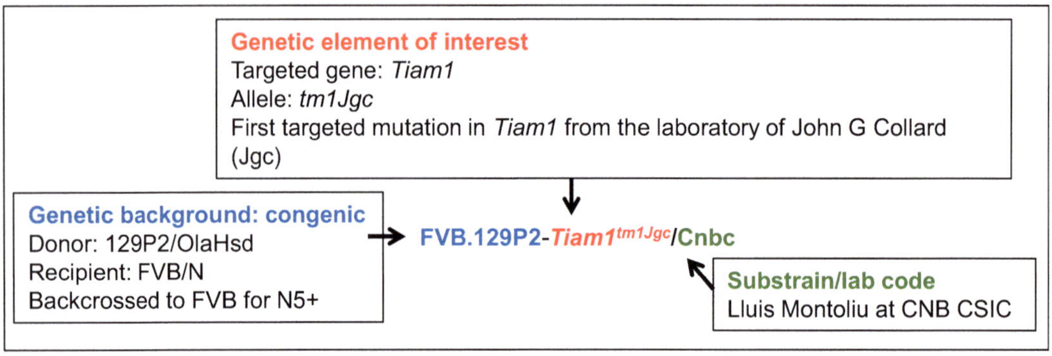

Fig. 6 Parts of a strain name—targeted mutation

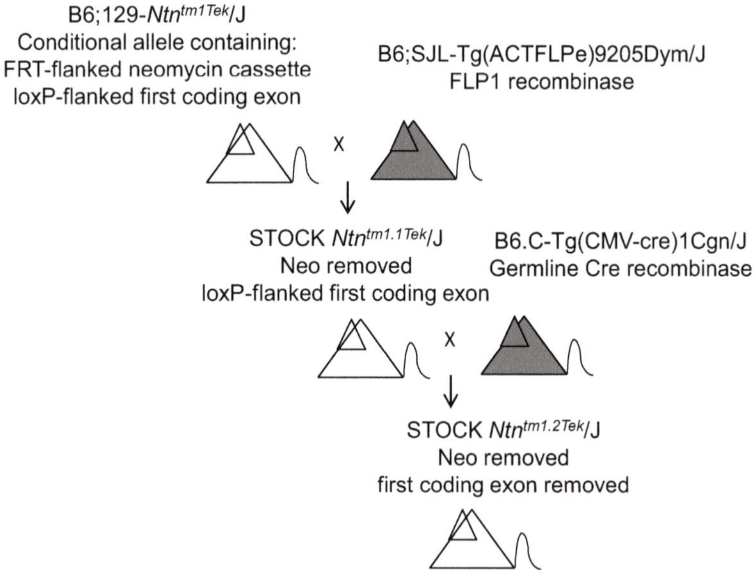

Fig. 7 Targeted mutation nomenclature

superscripted to the gene symbol, for example, *Trp53^{tm1Tyj}*. Notice that both gene and allele symbols are italicized in Fig. 6.

The first targeted mutations were created by inserting a neomycin resistance gene cassette into the gene of interest to disrupt transcription. These are known as knockouts (KOs). Over time, targeting cassettes have become significantly more complex and may include elements (*loxP* or FRT sites) that allow the conditional expression of the targeted gene and/or elements within the targeting construct. The first targeting event is labeled *tm1*. Subsequent heritable modifications of the initial targeted allele (for example, via *cre*- or *flp*-mediated recombination) are indicated by a decimal point and sequential numbers following the original allele name (for example, *tm1.1*, *tm1.2*) (Fig. 7). Note that the genetic background changes as new strain contributions are introduced by crosses to flp or cre recombinase-expressing strains. A second targeting event in the same gene from the same laboratory would be designated *tm2*.

Targeted mutations that are generated by the insertion of a foreign gene into the target gene are designated by inserting the gene symbol in parenthesis between the *tm#* and the lab code. $Lyz2^{tm1(cre)Ifo}$ is a knockin mutation in the which a cre recombinase gene is inserted into *Lyz2*. In this case, the endogenous promoter directs expression of the introduced gene.

A final mention must be made of alleles generated through large-scale production of genetic mutations in the mouse, in particular, the IKMC (International Knockout Mouse Project Consortium). Mice from these projects include a project abbreviation by inserting the project name in parenthesis between the *tm#* and the lab code: $Angptl8^{tm1(KOMP)Mbp}$ and $Tspo^{tm1b(EUCOMM)Wtsi}$ are two examples, the first created by KOMP (Knockout Mouse Project), the second by EUCOMM (European Conditional Mouse Mutagenesis Program). In addition, the IKMC targeting strategies utilize the designations *tm1a*, *tm1b*, *tm1c*, and *tm1d* for their derivative alleles. These strategies are described on the International Mouse Phenotyping Consortium (IMPC) website [11].

4.2 Endonuclease-Mediated Mutations

Endonuclease-mediated mutations (ZFN, TALEN, and CRISPR/Cas9) are generated using a technique that employs an element that recognizes and binds specific DNA sequences (zinc finger proteins, TAL effector proteins, and guide RNA, respectively) and a nuclease (Fokl, Fokl, and Cas9, respectively) that induces cleavage of double-stranded DNA to allow gene editing by endogenous homology-directed repair (HDR) or nonhomologous end-joining repair (NHEJR) [12, 13].

Unlike transgenic and targeted mutations, to which certain inbred strains are more amenable (FVB/N or C57BL/6 oocytes, and 129–derived ES cells, respectively), modification using ZFN, CRISPR, and TALEN can be performed using any strain, including strains that carry other mutations. One complication of the technique is that it can generate multiple off-target mutations in each "founder" animal due to random nucleotide insertion/deletion during NHEJR [14]. As a result, a formal allele name is not assigned until the desired mutation from a single founder has been molecularly defined.

The nomenclature for endonuclease-mediated mutations is similar to that for targeted mutations, but uses "*em*" (endonuclease-mediated mutation) rather than *tm* (targeted mutation) followed by a number and lab code, all italicized and superscripted to the gene symbol—for example, $Dolk^{em1J}$. As with targeted mutations, if multiple mutations are made in the same gene by the same lab, they are numbered sequentially (e.g., *Dicer*^{em1Psv}, *Dicer*^{em2Psv}, *Dicer*^{em3Psv}).

4.3 Gene Trap Mutations

Gene trapping is a strategy for insertional mutagenesis that uses a promoterless reporter gene (usually βgeo) inserted into either an intron or an exon of the "trapped" gene resulting in the expression of reporter from the endogenous promoter and reduced expression or inactivation of the trapped gene. The gene trap becomes an allele of the trapped gene and is designated as *Gt(vector or mutant ES cell line)Labcode*—for example, *Adamts4*$^{Gt(OST106997)Lex}$. Most gene traps are produced by one of several high throughput gene trap pipelines [15].

4.4 Chemically Induced Mutations

A chemically induced allele is designated by an abbreviation for the mutagen (e.g., *enu*) or by "*m*" (mutation) followed by a serial number and laboratory code (e.g., *m1Pas*), or by the initials of the mutagenesis center and a number (e.g., *Hlb301*). "Hlb" refers to the *H*eart, *L*ung and *B*lood mutagenesis program at The Jackson Laboratory. Initially, the gene is unknown and the allele and gene designations are the same. Once the gene is identified, the initial designation is superscripted to the gene symbol. For example, the strain name C57BL/6 J-*Hlb301*/J would change to C57BL/6 J-*Ldlr*^{Hlb301}/J upon identification of a causative gene (in this case *Ldlr*).

A dominant allele produces a phenotype whether it is present in one or two copies (heterozygous or homozygous, respectively). In the nomenclature, the first letter of a dominant allele is capitalized. As Hlb301 is a dominant allele, it is designated *Ldlr*^{Hlb301}. A recessive allele produces a phenotype only if it is present in two copies (homozygous). In the nomenclature, the first letter of a recessive allele is lower case. If the allele in the example were recessive it would be *Ldlr*^{hlb301}. Most alleles are recessive.

4.5 Transgenes

Transgenic mice carry a segment of DNA (foreign or mouse) incorporated into their genome via nonhomologous recombination. DNA is introduced into cells of the early mouse embryo by pronuclear microinjection or infection with a retroviral vector. Insertion of the DNA construct into the recipient genome is random and integration of multiple copies of the transgene usually occurs at a single site in a head to tail array. The new gene is expressed from a promoter that is part of the transgenic construct. In most cases, new DNA results in a gain of function either by expression of a new protein, or an altered protein, or by overexpression an existing protein. Usually, multiple transgenic founder mice are created and designated by using letters or numbers. In most cases, only a few founders are bred and maintained. It is important to note that there can be significant differences between

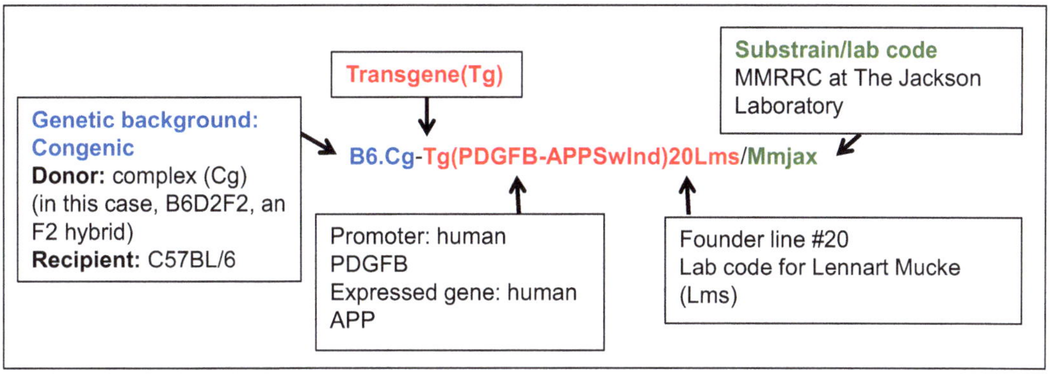

Fig. 8 Parts of a strain name—transgene

founder lines. Two strains that have otherwise identical nomenclature, but with different founder line numbers, are not necessarily interchangeable.

In the nomenclature, a transgene is designated Tg(insertion)#Labcode; details about the transgenic construct are within the parenthesis. In Fig. 8, the information within the parentheses () includes the gene symbol of the promoter, followed by a hyphen, and the gene symbol of the expressed gene (Fig. 8). Notice that the gene symbols inside the parenthesis are entirely capitalized, indicating that they are human genes. Neither the entire transgene name nor the gene symbols within the parenthesis are italicized. The nomenclature guidelines for designating information inside the parentheses are complex and will not be addressed here. The full set of rules and guidelines are available through the Mouse Genome Database [2].

5 Conclusion

This chapter focuses on the basics of mouse nomenclature and the nomenclature of genetically engineered mutations. There are a variety of other types of strains and mutations not presented here that have unique nomenclature. Comprehensive information is available through the Mouse Genome Database [2], as well as numerous texts such as *The Jackson Laboratory Handbook on Genetically Standardized Mice* [16], Lee Silver's *Mouse Genetics* [10], and training courses sponsored by the American Association of Laboratory Animal Science (www.alaas.org), The Jackson Laboratory (www.jax.org), and Charles River Laboratories (www.criver.com).

References

1. Dunn LC, Gruneberg H, Snell GD (1940) Report of the committee on mouse genetics nomenclature. J Hered 31:505–550
2. International Committee on Standardized Nomenclature for Mice (2014) Rules and guidelines for nomenclature of mouse and rat strains. Mouse genome informatics. The Jackson Laboratory, Bar Harbor, Maine. http://www.informatics.jax.org
3. Davisson MT (1996) Rules for nomenclature of inbred strains. In: Lyon MF, Rastan S, Brown SDM (eds) Genetic variants and strains of the laboratory mouse. Oxford University Press, Oxford, pp 1532–1536
4. Festing MFW (1999) Inbred strains of mice. Mouse genome informatics. The Jackson Laboratory, Bar Harbor, Maine. http://www.informatics.jax.org
5. Eppig JT, Strivens M (1999) Finding a mouse: the International Mouse Strain Resource (IMSR). Trends Genet 15:81–82
6. Mouse phenome database web site (2016) The Jackson Laboratory, Bar Harbor, Maine. http://www.jax.org/phenome
7. Petkov PM, Ding Y, Cassell MA, Zhang W, Wagner G, Sargent E, Asquith S, Crew V, Johnson KA, Robinson P, Scott VE, Wiles MV (2004) An efficient SNP system for mouse genome scanning and elucidating strain relationships. Genome Res 14(9):1806–1811
8. Institute for Laboratory Animal Research (ILAR) (2016) U.S. National Academy of Sciences Washington DC. http://dels.nas.edu/ilar
9. Chia R, Achilli F, Festing MFW, Fisher EMC (2005) The origins and uses of mouse outbred stocks. Nat Genet 37(11):1181–1186
10. Silver LM (1995) Mouse genetics. Oxford University Press, New York
11. International Mouse Phenotyping Consortium (2016). http://www.mousephenotype.org/about-ikmc/targeting-strategies
12. Gaj T, Gersbach CA, Barbas CF (2014) ZFN, TALEN and CRISPR/Cas-based methods for genome engineering. Trends Biotechnol 31(7):397–405
13. Sander JD, Joung JK (2014) CRISPR-Cas systems for editing, regulating and targeting genomes. Nat Biotechnol 32:347–355
14. Fu Y, Foden JA, Khayter C, Maeder ML, Reyon D, Joung JK, Sander JK (2013) High frequency off-target mutagenesis induced by CRISPR-Cas nucleases in human cells. Nat Biotechnol 31(9):822–826
15. International Gene Trap Consortium (2016). http://www.genetrap.org/
16. Flurkey K, Currer J (eds) (2009) The Jackson Laboratory handbook on genetically standardized mice. The Jackson Laboratory, Bar Harbor, Maine

Chapter 14

Breeding Strategies for Genetically Modified Mice

Jan Parker-Thornburg

Abstract

Genetically modified mice can be generated using a variety of methods. Depending on the method used, the breeding strategy must be modified not only for the initial founding generation of mouse, but to establish individual mouse lines as well. Transgenic founder mice are each unique and must be outcrossed to initially establish the line. Mice that have been targeted using embryonic stem cells will need to be tested for germ line transmission of the modified gene prior to establishing a line. Mice that have been targeted using CRISPR/Cas9 gene-editing endonucleases will often be mosaic, and thus, also require testing for germ line transmission of the mutation of interest.

Key words Transgenic, Knock-out mice, Knock-in mice, CRISPR, CRISPR/Cas9, Germ line transmission, Chimera, Mosaic

1 Introduction

There are three major methods currently used to generate genetically modified (GM) mice. GM mice can be created by classical transgenesis [1], by gene targeting in ES cells [2] or by gene editing using various gene-editing endonucleases such as CRISPR/Cas9 [3, 4]. The method of establishing a mouse line differs for each of these, and therefore, there are several questions that should be answered when devising breeding strategies for genetically modified mice. First, what type of model was generated? Transgenic? Knock-out? Knock-in? Next, how does one establish the line from the initial animals produced? Finally, how does one maintain the line or lines after establishing them? All of these should be considered when setting up the breeding strategy.

1.1 Classical Transgenesis

In classical transgenesis, transgenic mice are created by injection of a DNA construct into the pronucleus of a fertilized egg [5]. The incorporation of the DNA is generally random, the DNA can integrate into several locations, and each site of integration can have anywhere from one to several hundred copies [6]. Because

Melissa A. Larson (ed.), *Transgenic Mouse: Methods and Protocols*, Methods in Molecular Biology, vol. 2066,
https://doi.org/10.1007/978-1-4939-9837-1_14,

of the random nature of the integration, the transgene in each founder—the F0 initial animal identified with the transgene—will have integrated into a different site in the genome. Each transgenic founder is unique, and should not be mated to another founder. To establish a line from a transgenic founder, s/he is mated to a wild-type animal of the same strain to obtain progeny with the exact mutation of the founder. If the gene is not detrimental, approximately 50% of the pups from this outcross will contain the gene. These F1 pups can then be tested for phenotype, crossed to animals with other mutations, and bred to maintain the line.

1.2 Classical Gene Targeting

Classical gene targeting is done using mouse embryonic stem (ES) cells, and the initial genotyping is done at the ES cell stage [2]. Correctly targeted cells from between one and three ES cell clones are injected into mouse blastocysts to generate chimeric mice (the resulting mouse has a mixture of cells from both from the mouse blastocyst and from the injected ES cells). Chimeras are generally identified by coat color differences, with the ES cells carrying genes for one coat color and the blastocyst carrying genes for another coat color, generally recessive to that of the ES cells [7]. (Alternatively, strain differences in key genes can be used for identification, but this requires PCR analysis and genotyping to determine which mutation/cell is present.) One mates the chimera back to a wild-type mouse of the same strain as the blastocyst to generate progeny. If the ES cells have contributed to the germ line, pups will be born that carry the coat color mutation of the ES cells, and we can say that the ES cells have exhibited "germ line transmission." One caveat is that the ES cell clones are generally heterozygous and have a diploid genome, whereas the germ line is haploid. Thus, it is equally expected that a mouse exhibiting germ line transmission of the ES cells could carry the wild-type or the mutant allele. This necessitates genotyping of all germ line-transmitted pups. Once the correct genotype is detected in the pups, they are then mated. Mating is often done initially with wild-type mice to generate additional pups of the correct genotype. However, at this time, one can also mate the germ line founder with Cre recombinase [8] or Flp recombinase [9] lines, with mice containing other mutations, or with other germ line founders from the same clone to establish homozygous lines.

1.3 CRISPR/Cas9 Gene Targeting

CRISPR/Cas9 gene targeting involves injection of a guide RNA, the Cas9 nuclease (either protein or mRNA) and, if generating an insertion or knock-in mutation, a donor DNA into the pronucleus of a fertilized egg [4]. Mice generated using CRISPR/Cas9 technology are often mosaics (a mixture of cells from the mouse, each which may have different mutations) due to the time over which Cas9 nuclease acts [10]. The mosaics are treated much like the chimeras generated in classical gene targeting. They are mated to wild-type animals to establish germ line transmission of each mutation that occurs. An important caveat is that one will often obtain multiple mutations from one G0 founder animal [10]. Coat color cannot be used to determine if there are germ line progeny, as all of the cells, even those with different mutations, arose from the original egg injected. Thus, each G1 pup can potentially be a founder, and germ line transmission must be detected by genotyping. Mutant pups can then be grown and mated similarly to those from classical gene targeting to establish the mutant lines.

2 Methods

2.1 Classical Transgenic Breeding

There are some considerations to bear in mind when establishing a breeding program for transgenic mice. The first generation after injection (the founder animals) are considered F0. The site of transgene integration is random, and the number of integrated copies can vary from one to hundreds. It is important to test more than one F0 line, as expression of the transgene can be affected by the position of chromosomal integration. Breeding schemes are as follows:

1. Outcross F0 founder animals to wild-type mice of your desired strain. For female founders, mate one female to one male. For male founders, mate one male to two females (*see* **Note 1**).
2. When mating for line maintenance, identify three or four founder lines that express the representative or desired phenotype. Initially, these lines should be outcrossed to wild-type animals. It is possible to mate hemizygotes together (within a line) to generate homozygous lines, but such matings may result in lethality or sterility (occurs in about 20% of transgenic lines).

2.2 Classical Gene Targeting Using ES Cells

After injecting correctly targeted embryonic stem cells into blastocysts, those embryos into which the ES cells have been incorporated will develop into a chimera (as indicated by coat color). Since the ES cells are male, most chimeras will be male, although in rare cases female chimeras are noted. There is only one site of integration and one copy integrated, as the modified cells are a result of targeted homologous recombination in the gene of interest. Although it is recommended to inject multiple clones, the chimeras between clones should be equivalent, provided the ES cells were correctly targeted. Mating schemes are as follows:

1. If available, test a minimum of five male chimeras per clone. If less than five male chimeras are available, or if all the male chimeras have a low percentage of ES cell contribution (as determined by coat color), mate female chimeras, too.

Backcross each chimera to a wild-type mouse of the same genetic background as the host blastocysts. To test for germ line transmission, mate each male chimera to at least five females. Check plugs daily, and once detected, remove plugged female(s) and replace with a new female (*see* **Note 2**). Mate each high coat color-percentage female chimera to one male in a continuous mating scheme to maximize pregnancies and the number of pups (*see* **Note 3**).

2. Once germ line transmission (glt) has been identified (by coat color and genotyping) (*see* **Note 4**), outcross males to both a wild-type female of the same strain as the ES cells and to a female germ line animal. This will enable propagation of both the line (wild-type mating) and homozygosity (germ line mating). Mate two or three germ line mice per clone.
3. Test expression or gene function in mice from two to three different clones. There should be no variation between them. If no variation is seen, reduce colony size to include the progeny from only one clone. If variation is seen, establish lines from representative mice of each differing clone and maintain them as sublines. Maintain two to three mating pairs for each clone. At this point, males and females from the same clone can be inbred to establish homozygous lines (if not performed earlier). Progeny from brother/sister matings should have approximately 25% homozygous for the targeted gene. If no homozygotes result or if many fewer than 25% of pups are homozygous, it is likely that homozygosity confers lethality or reduced viability, and the line should be maintained in the heterozygous state. Some instances of lethality or sub-viability can be rescued by mating to another genetic background (e.g., crossing a gene that's lethal in C57BL/6 into a BALB/c background).
4. Once homozygous lines are generated, maintain three to five mating pair at all times of staggered ages (to prevent requiring replacement at the same time). In addition, it is advisable to cryopreserve sperm or embryos from the heterozygous line as a backup in case of later line loss or contamination (*see* **Note 5**).

2.3 Gene Targeting Using CRISPR/Cas9 Technology

While CRISPR/Cas9 technology represents a rapid system for the generation of gene-edited mice, the resulting mice and subsequent genotyping are not so straightforward. As such, the first generation of founder animals is considered G0 and is often mosaic (i.e., the mouse may have cells with different mutations). Although Cas9 may cut, the DNA could be exactly repaired so that no mutation is seen, or there may be small indels (insertions or deletions). It is also possible for homology driven repair to insert a sequence, if an exogenous template is provided. In addition, there is the possibility that off-target sites could be affected. These considerations dictate the mating scheme for CRISPR mice.

1. If looking for homology driven repair (insertion), mate all mice (even those that appear wild-type) (*see* **Note 6**). As the mice are likely mosaic, they may transmit the correct gene through the germ line even if it is not detected by genotyping. To test for germ line transmission, backcross G0 animals to wild-type mice. Mate each male to at least five females; check plugs daily and remove plugged females to a separate cage, and replace plugged female with a new mouse. Mate each G0 female to one male in a continuous mating scheme to maximize pregnancies and pups.
2. Once germ line transmission has been identified, outcross glt progeny to wild-type mice of the same strain. Mate all mice that exhibit the correct genotype; one male to two females, and one female to one male.
3. As different G1 mice may have slightly different mutations, it is advisable to keep mice with genotypes that could be instructive. Establish lines from representative mice within each different mutation and maintain these as sublines. Mate males one to two with wild-type females, and mate G1 females one to one with wild-type males. Genotype progeny by PCR and sequencing, as was performed for the G0 mice. Maintain two to three mating pair per subline.
4. Once each subline is established, males and females can be inbred to generate homozygous lines. Set up matings 1:1 with brothers and sisters containing the same genotype. Progeny from the brother/sister matings should be approximately 25% homozygous for the targeted gene. If no homozygotes are seen, or if many fewer than 25% progeny are homozygous, then it is likely that homozygosity confers lethality or sub-viability and the line should be maintained as heterozygous. In instances of lethality or sub-viability, rescue is sometimes possible by crossing the gene into another genetic background (e.g., crossing a gene that is lethal in C57BL/6 into a BALB/c background). One homozygous lines are established, maintain three to five mating pair of mixed ages, to avoid requiring replacement at the same time (*see* **Notes 7** and **8**).

3 Notes - Common Errors

1. Mating F0 transgenic founders to each other. As each F0 is unique, it is essential to mate the F0 to wild-type to "fix" the gene in the germ line.

2. Not obtaining sufficient numbers of progeny from each chimera in a gene targeting experiment; a minimum of 50 progeny should be obtained for EACH chimera. Especially with less optimal ES cells, contribution to the germ line may be minimal, requiring many more pups to find it.
3. Failure to mate high percentage female chimeras derived from gene targeting experiments. Female chimeras can be XO, and, while less fertile, can often still give progeny, some of which may contain the gene of interest.
4. Not genotyping targeted mice that are identified as glt based on coat color. Unless care was taken to generate a homozygous ES cell line, only half of glt mice can be expected to have the targeted gene.
5. Failure to maintain a "stock" line with the original mutation when outcrossing to add in additional genes. To be safe in case of a breeding depression or loss of a line, it is important to maintain the original stock line—either live or frozen.
6. Presuming that G0 genotyping is definitive in CRISPR/Cas9 experiments. All G0 animals should be mated and G1 progeny tested for genotype.
7. Delaying mating until the animals have passed prime sexual maturity (3–7 months of age).
8. Not using optimal pairings: 1:1 for females of interest, 1:2 or 1:3 for males of interest. When mating males, using too few females will result in fewer progeny. Using too many females can occasionally lead to the death of the male.

Acknowledgments

This work was supported by NCI grant #CA016672 (GEMF), a Cancer Center Support Grant to MD Anderson Cancer Center and NCI grant #R50CA211121.

References

1. Behringer R (ed) (2014) Manipulating the mouse embryo: a laboratory manual, 4th edn. Cold Spring Harbor Laboratory Press, New York
2. Joyner AL (ed) (2001) Gene targeting: a practical approach, 2nd edn. Oxford University Press, Oxford
3. Wang H, Yang H, Chivalila CS, Dawlaty MM, Cheng AW, Zheng F, Jaenisch R (2013) One-step generation of mice carrying mutations in multiple genes by CRISPR/Cas-mediated genome engineering. Cell 153:910–918. https://doi.org/10.1016/j.cell.2013.04.025
4. Singh P, Schimenti JC, Bolcun-Filas E (2015) A mouse geneticist's practical guide to CRISPR applications. Genetics 199:1–15. https://doi.org/10.1534/genetics.114.169771
5. Brinster RL, Chen HY, Trumbauer M, Senear AW, Warren R, Palmiter RD (1981) Somatic expression of herpes thymidine kinase in mice following injection of a fusion gene into eggs. Cell 27:223–231

6. Brinster RL, Chen HY, Warren R, Sarthy A, Palmiter RD (1982) Regulation of metallothionein--thymidine kinase fusion plasmids injected into mouse eggs. Nature 296:39–42
7. Williams E, Auerbach W, DeChiara TM, Gertsenstein M (2011) Combining ES cells with embryos. In: Pease S, Saunders TL (eds) Advanced protocols for animal transgenesis. Springer, London
8. Nagy A (2000) Cre recombinase: the universal reagent for genome tailoring. Genesis 26:99–109
9. Rodriguez CI, Buchholz F, Galloway J, Sequerra R, Kasper J, Ayala R, Stewart AF, Dymecki SM (2000) High-efficiency deleter mice show that FLPe is an alternative to Cre-loxP. Nat Genet 25:139–140. https://doi.org/10.1038/75973
10. Yen ST, Zhang M, Deng JM, Usman SJ, Smith CN, Parker-Thornburg J, Swinton PG, Martin JF, Behringer RR (2014) Somatic mosaicism and allele complexity induced by CRISPR/Cas9 RNA injections in mouse zygotes. Dev Biol 393:3–9. https://doi.org/10.1016/j.ydbio.2014.06.017

Chapter 15

Strategies for Behaviorally Phenotyping the Transgenic Mouse

Kenneth E. McCarson

Abstract

The techniques and protocols to modify the mouse genome described in this volume allow researchers to produce genetic models of a remarkable number and breadth of human disease. The generation of gene-modified mice offers profoundly powerful approaches for bringing known or purported human gene disruptions into mouse models, but the degree to which the resultant mutant mouse recapitulates the complex physiological and behavioral features of the human disease state is a key variable in the ultimate usefulness of the mouse model organism. Accordingly, the behavioral characterization of mice with novel targeted gene mutations is an important initial step in determining the potential impact of a novel mouse model. This chapter addresses strategies useful in the initial observations of the animal that assist in directing the choice of secondary tests to assess more detailed aspects of potentially disrupted behaviors that may be relevant to the disease being modeled. An initial standardized, comprehensive screen that assesses general health, reflexes, and sensorimotor functions is the first step in characterizing behavioral phenotype, and results often suggest areas where more complex complementary behavioral assays may reveal more detailed disruption of normal behavior. This sequential, standardized approach reduces variability between subjects; this chapter also addresses approaches to reducing experimental artifacts due to handling, test order, testing facility environment, and other sources. This brief overview of behavioral phenotyping approaches is intended to provide practical information to streamline initial characterization of new mouse models and maximize the usefulness of efforts to use these models to study human health and disease.

Key words Behavior, SHIRPA, Screening, Locomotion, Sensory function, Cognition, Phenotype

1 Introduction

Mutant mice can be useful surrogates for understanding human genetic disorders. Humans and mice share significant portions of their genomes, and large numbers of orthologous genes can be edited to faithfully recreate human mutations or to produce useful alterations in normal physiology to mimic human disease. While models based on simpler organisms exist with potentially higher throughput of data collection, the number of common genes is fewer, thus limiting the number of disorders that can be studied.

Melissa A. Larson (ed.), *Transgenic Mouse: Methods and Protocols*, Methods in Molecular Biology, vol. 2066, https://doi.org/10.1007/978-1-4939-9837-1_15,

Even when simpler models are available, discoveries typically must be confirmed in mice where complex behavioral phenotypes can be characterized. Furthermore, the rapidity with which mice can be generated with current gene editing technologies reduces differences in throughput and expense.

1.1 Mutant Mouse Strains

The production of mouse models using the approaches described in this volume can provide genotyped stocks of wild-type and mutant mice in numbers adequate for behavioral phenotyping. In mouse models of human disease, if a genetic variant or environmental perturbation (typically neurological) is responsible for behavioral dysfunction, then a mouse incorporating the variant or perturbation should display corresponding behavioral changes. The use of standardized, robust behavioral analyses of the mutant mouse can assess whether specific alterations lead to disease-relevant phenotypes. Many disorders in humans, particularly those of the nervous system, are associated with deficits in cognitive, sensory, motor, or social aspects of behavior. Behavioral screening using a battery of well-defined tests can help determine whether the mutant mice display behavioral features similar to the human disease.

Selection of the background strain used for development of the mutant mouse model is a significant issue when plans include eventual phenotyping of even simple behavioral traits. Inherent strain variations occur that produce abnormalities that can likely compromise the mouse in future behavioral tests. The most confounding issues arise in background strains that spontaneously develop blindness, deafness, or other physical defects that limit sensory or locomotor function. For example, age-related hearing loss and vestibular defects are associated with many (129) strains, A/J, C57BR, C57L, DBA, LP, NOD, ALR, ALS, and C57BL/6J. In some of these strains vision loss occurs early; in C57BL/6J it occurs as late as 18 months of age. Similarly, blindness predictably appears in FVB, C3H, BUB, CBA, SJL, SWR, and NON strains. Despite the popularity of some strains for transgenic modification techniques, such as the FVB strain with large embryo and litter sizes, known sensory deficits can render these strains useless in the context of behavioral phenotyping. Thus, care should be taken to use the foresight needed to develop mutant mouse models in strains compatible with behavioral phenotyping tests that will remain viable for the needed duration of study. In the course of cross-breeding mutant strains to create multiple mutation models, mixed background strains may emerge. This may be advantageous if known physical defects in a founder strain can be reduced or avoided by mixing the genetic background, but care should be taken to ensure that comparisons are made across consistent background strains.

1.2 Sex Differences

It has long been assumed that the estrous cycle of female rodents results in an added source of variability in experimental data that could appropriately be avoided by studying only male subjects. Despite mandates from funding agencies that both male and female subjects be included in all studies, there continues to be an ongoing robust bias toward studies using only male subjects. Most experimental endpoints do not show robust sex differences, yet the use of female mice in phenotyping strategies has been a rare practice because of the perceived variability related to estrous cycling.

The study of sex differences is an important and growing field of exploration, with notable sex differences apparent in areas of simple behavior, such as pain sensation [1]. When specifically addressed, sex differences can be appropriately studied using both male and female mice to determine whether sex differences exist in the mutant model being tested. Specific guidelines exist for the optimal experimental procedure to determine sex differences in behavioral assays [2]. Initial screening stages of any study should be to compare gonadally intact adult males and females; subsequently, cycling females, ovariectomized females, and hormone replacement can follow to gather more details about the role of sex hormones in modifying a given behavioral outcome.

Nonetheless, good evidence supports the idea that strain- or mutation-related differences in behavior are often more robust than any underlying sex differences. For example, in assays of open field locomotion, tail-flick and tail-suspension tests, Rotarod, startle reflex and prepulse inhibition, and tail-flick and hot-plate thermal withdrawal tests, BALB/cByJ (with high sex variability) and C57BL/6J (with no sex variability) showed consistent strain-related differences regardless of the sex of subjects [3]. These types of findings undermine long-held assumption of high estrous cycle variability of data from female mice [4]. In the initial behavioral screening of novel lines of mutant mice, numbers of experimental subjects are often limited, and the benefit of testing available mice of both sexes clearly outweighs the potential confounds of estrous cycle effects on data variability.

1.3 Organization of Testing

In addition to strain and sex, many features of a phenotyping strategy can have an impact on the quality of results. Ultimately, sound experimental design is of the most critical importance in avoiding unwanted confounds that can introduce unknown sources of variability that can seriously compromise the resulting data. The use of adequate and appropriate numbers of experimental subjects with planned, correct statistical analysis methods are, as always, of primary importance in designing the phenotyping strategy. However, with mutant mouse colonies of limited size it can be impractical or even impossible to generate age-matched cohorts of equal numbers of male and female test subjects in each of the relevant treatment groups to facilitate simultaneous assessment of specific behavioral endpoints. Even if this is possible, the throughput available in several behavioral assays (e.g., days-long memory acquisition and retention trials) simply cannot accommodate very large numbers of experimental subjects without remarkable duplication of equipment or technicians. Accordingly, the use of well-planned, consistently timed testing paradigms allows multiple, sequential groups of mutant and wild-type mice to be phenotyped as progeny emerge in litter groups from even small colonies. By testing with consistent batteries of tests in age-matched cadres of subjects, the appropriate numbers of experimental subjects can be accumulated over multiple iterations of behavioral tests. This also provides another level of experimental randomization that can help minimize small sources of variability related to handling or testing order. Within this approach, however, it is critical that the representatives of various genotypes, treatments, and sex be included in each cadre, preferably in multiples of each combination, throughout the testing paradigm. This kind of approach will facilitate the most meaningful comparisons across wild-type, heterozygous, and homozygous genotypes while also preserving appropriate experimental controls for treatment and sex of subjects [5].

The total numbers of experimental subjects should be driven, if at all possible, by meaningful statistical power analyses of the subjects and tests to be used. Even in pilot tests of novel mutant strains of mice, initial estimates of statistical power needs can be made using proxy data from similar studies using similar behavioral assays. Complex behavioral assays can often require numbers of subjects that can seem astoundingly large to investigators used to biochemical or simpler physiological endpoints; it is not unusual for behavioral assays to require 12–20 subjects per treatment condition to provide adequate statistical power. Knowledge of these features of behavioral testing before phenotyping begins can help to demystify the relatively heavy labor and resource demands of behavioral testing for investigators new to the arena of behavioral assays.

In order to maintain adequate organization of the resulting complex sets of experimental groups, accurate identification methods can be critical in maintaining the identity of individual experimental subjects. Approaches such as toe clipping, ear punches, ink tattooing, or subcutaneous ID transponders can be useful in this regard [6]. Of course, meticulous record keeping underlying the process is critical. Housing of weaned, genotyped mutant mice should not be segregated by genotype; ultimately, representatives of all genotypes and treatments in an experiment should be housed as randomly as possible when distributed across cages of 3–5 individuals. As mice are social animals, individual housing should be avoided to every extent possible and environmental enrichment provided. Subjects entering the phenotyping process should be as closely age matched as is practical or possible.

1.4 Behavioral Phenotyping Facilities

Ideally, testing can be performed within the animal housing facility or in procedure rooms nearby—although segregation of behavioral testing equipment from the harsh cleaning conditions of rodent housing rooms can be an important aspect of maintaining their repair in the long term. In any event, the transport of subjects from the home-cage housing environment to the testing facility should be as short and gentle as possible. The stressors of transport can be minimized by providing adequate time between transport and testing to allow the mice to acclimate to their new surroundings. This should be a minimum of 15–20 min, but longer times are better if the experimental protocol allows. Generally speaking, groups of experimental subjects should not be held in the immediate testing area; some separation and distance between subjects being currently tested and those previously tested or "in line" for testing should be maintained if possible.

Every effort should be taken to adequately blind the observing experimenter from the genotype and treatment group of the subjects being phenotyped. While it is often impossible to completely blind the observer due to observable genotype-related features (e.g., coat color or body weight), attention to this issue is a key feature of reducing the impact of any observer bias on the data gathered. In this regard, it is often useful to have some consistent division of labor among investigators in behavioral phenotyping, where one individual might mark, record, and transport animals, while another performs the actual behavioral assay. This approach can have the added benefit of reducing interobserver variability in data collection within a single behavioral assay.

Well-designed behavioral testing suites in animal housing facilities [7] and rigidly standardized behavioral phenotyping strategies can provide a means to gather useful behavioral information about mutant mice. Reducing procedural variability helps maximize the replicability and impact of mouse behavioral phenotyping to gather information about mutant genotypes that provide important tools to translate preclinical hypotheses into developing treatments [8].

2 Brief Descriptions of Commonly Used Tests

2.1 Appearance and Overall Health

Even the most complex phenotyping paradigms begin with simple assessments of visually observable or easily measured features of the mutant mouse. When well organized, an initial screen of novel mouse genotypes can provide substantial information about where to focus subsequent behavioral testing. Numerous teams of investigators have used various batteries of tests for screening behavioral disruptions in mice; the tests involve various degrees of breadth and complexity [9–13]. By far, the most widely used, validated, and comprehensive approach is the SHIRPA protocol. The acronym SHIRPA represents SmithKline Beecham, Harwell,

Imperial College, Royal London Hospital Phenotype Assessment, initially proposed in 1997 by investigators from the named institutions [12, 14]. The screening paradigm uses three stages of tests of increasing complexity designed to mimic human neurological and psychiatric clinical diagnostic procedures. Modified SHIRPA approaches include up to 40 tests of increasing focus and complexity [14]. The second and third stages of SHIRPA testing involve very detailed assays of behaviors related to neurological deficits and analysis of potential underlying pathologies [15, 16]; these stages are less widely used, however, than the initial phase of the SHIRPA battery, which comprised a series of simple tests that provide an excellent means to obtain a relatively broad initial screen of mutant mouse phenotype [12, 17]. It is powerful in providing a very informative, high-throughput approach for observation of phenotypic traits and includes quantitative features allowing scoring of both normal and unanticipated overt phenotypes. Subjects are tested individually and sequentially, and the initial battery of observations only takes about 5 min per subject. Over time, the initial SHIRPA screen has been modified to include more observations of morphology [18], but still retains the essential features of the original protocol (Table 1).

2.2 Body Weight

Mutation of genes in the mouse often results in abnormal body weight phenotypes. Since numerous other biological traits vary as a function of total body weight, this can be a confounding factor in interpreting genotype–phenotype associations. Correlation of body weight or total body length with some endpoints (e.g., gait dynamics or Rotarod performance) can produce misleading interpretation of the source of phenotypic differences. Rigorous statistical modeling that accounts for body weight can help avoid spurious interpretations of gene mutations [19].

2.3 Body Temperature, Food, Water, Metabolism, and Exercise

Changes in locomotor phenotype can reflect alterations in the basal metabolism of mutant mice. However, there may also be more subtle metabolic manifestations related to the level of hypo- or hyperactivity associated with the mouse phenotype. Typically, only extremely large perturbations of metabolism will be manifested as measurably abnormal body temperatures or body weight. If more subtle metabolic changes are suspected, metabolic chamber testing can reveal quantitative details about the mouse metabolism (e.g., Promethion Indirect Calorimetry system, Sable Sytems International, Las Vegas, NV). Systems are available that continually measure oxygen consumption, carbon dioxide production, food and water intake/behavior, body mass, spontaneous physical activity, and wheel running (exercise). Body composition analysis of fat body mass, lean body mass, and water mass can be directly

Table 1
SHIRPA-based initial screening of basic mouse behaviors

	Parameter	Range of scores
Observation in transparent cylinder	Body position	0 = Completely flat to 5 = repeated leaping
	Spontaneous activity	0 = None to 4 = constant, robust movements
	Respiratory rate	0 = Slow and irregular to 3 = hyperventilation
	Tremor	0 = None to 2 = marked tremor
	Urination	Number of episodes
	Defecation	Number of episodes
Observation in locomotor arena	Transfer arousal	0 = Coma to 6 = extremely agitated
	Locomotor activity	= Number of grid squares crossed in 30 s
	Palpebral closure	0 = Eyes wide open to 2 = eyes completely shut
	Piloerection	0 = Absent, 1 = present
	Startle response	Magnitude of jerk response to sound (loud click); 0 = none, 3 = robust
	Gait	0 = Normal fluid gait to 3 = locomotor incapacity
	Pelvic elevation	0 = Flattened to 3 = elevated more than 3 mm
	Tail elevation	0 = Flaccid to 2 = erect (Straub) tail
	Touch escape	0 = None to 3 = vigorous withdrawal
	Positional passivity	When held by tail, 0 = struggles to 4 = no struggling
Observations during tail suspension	Abnormal behavior	0 = Absent, 1 = present
	Visual placement	Forepaw paddling; 0 = none to 4 = initiation greater than 22 mm from target
	Grip strength	0 = None to 4 = strong
	Body tone	Resistance to compression; 0 = flaccid, 2 = very firm
	Pinna reflex	Ear flick in response to light touch; 0 = none to 2 = pronounced
	Corneal reflex	Blink response to light touch; 0 = none to 2 = pronounced
	Toe pinch	0 = None to 4 = very brisk withdrawal
	Hanging grasp	0 = Active and grips with hind paws to 4 = falls immediately
Observations during supine restraint	Body length	Measured in mm (nose to base of tail)
	Skin color	0 = Blanched to 2 = deep red plantar surface
	Heart rate	0 = Bradycardia to 2 = tachycardia
	Limb tone	0 = No resistance to 2 = extreme resistance
	Abdominal tone	0 = Flaccid to 2 = extreme resistance
	Lacrimation	0 = Absent, 1 = present
	Salivation	Wetness of fur under chin; 0 = none to 2 = extreme
	Provoked biting	0 = Absent, 1 = present
Observations in nonspecific test area	Righting reflex (RR)	0 = No impairment (rapid righting) to 3 = fails to right when placed on back
	Contact RR	0 = Absent, 1 = present
	Negative geotaxis	Movement on vertical hanging screen; 0 = none, 4 = rapid climbing
	Fear	0 = Absent, 1 = present
	Irritability	0 = Absent, 1 = present
	Aggression	0 = Absent, 1 = present
	Vocalization	0 = Absent, 1 = present
	Body temperature	Measured in degrees centigrade

measured through imaging techniques (e.g., EchoMRI-100 body composition analyzer). The data collection and analysis for these measurements can be complex, but can reveal otherwise subtle phenotypes associated with genetic disruptions.

2.4 Tests of Social Interaction

Subtle changes in a mouse's cognitive function, olfaction, or vision can produce aberrant behaviors that can manifest in the way that individual interact with other individuals or groups of mice. Several behavioral testing approaches can reveal details about how a mutant mouse's social behaviors are different from wild-type mice. These assays are based on well-established models of rodent social interactions [20–24].

If group housed in genotype-mixed caging, simple observation of home-cage behaviors by an observer not blinded to genotype can suggest social interaction issues that can be studied in greater detail.

2.4.1 Social Interaction Arenas

Social approach/avoidance testing paradigms are useful in assessing the interactions between a test subject and a familiar, novel, or juvenile conspecific mouse. This is typically accomplished using a three-chambered arena [25] with a chamber at one end containing the conspecific mouse. Briefly, a test subject is allowed to freely explore the testing arena for 5 min in the absence of other animals. After the exploratory period, a novel animal is placed into a perforated acrylic or wire mesh cylinder located at one end of the arena. An identical cylinder is left empty on the opposite side of the arena. The cylinders contain small holes through which the animals may sniff, touch minimally, and interact. The subject animal is allowed to explore the arena for an additional 10 min in the presence of the social stimulus. The entire session is either directly observed or video recorded and scored for time spent in each section of the arena, nose pokes, time spent within close proximity to the social stimulus, and other parameters as needed. Typically, young wild-type mice will spend more time in a chamber containing a stranger mouse than in an empty chamber (sociability) and will prefer to spend more time with an unfamiliar stranger than a more familiar conspecific (preference for social novelty) [8, 22]. Using similar approaches, social learning can be measured using repeat exposures to the arena and conspecific mice. These kinds of tests can be important in mouse models of schizophrenia and autism, which incorporate analogies to symptoms of autism such as abnormal social interactions, deficits in communication, and high levels of repetitive behaviors [26–28].

2.4.2 Olfactory Dishabituation

Another less complex measure of attention to social cues is the test of olfactory dishabituation, which measures the ability of a test subject to focus on novel social cues such as a glass slide with an anogenital swipe from another mouse. Nonsocial olfactory cues serve as an important control in this paradigm, where the cue is usually presented on the wall of an open field arena, and the test mouse observed for the proportion of time in the arena spent attending to the olfactory cue versus other areas of the arena.

2.4.3 Resident–Intruder Aggression Test

A more direct assessment of overtly aggressive behaviors, particularly in male mice, is the resident–intruder test, where a novel "intruder" mouse is introduced into the home cage of an acclimated, solitary housed test mouse. Latency to, numbers, and duration of aggressive interactions (chasing, biting, etc.) are then quantified over a 5- to 10-min exposure time by direct observation or scoring of video recording. Similar tests of aggressive behavior between dams of new litters and unfamiliar male mice offer some degree of analogous testing for female mice, but the tests are not directly comparable.

2.5 Tests of Sensory Function

Basic neurologic function, including sensitivity to auditory and sensory stimuli, can be assessed using tests revealing sensorimotor impairment or idiosyncratic responses to sensory stimuli. Basic orienting reflexes are tested as part of the initial SHIRPA screening (response to audible startle sound, pinna reflex, visual placement tests). Normal sight can be assessed using a very simplified version of the Morris water maze (see below), in which the target platform is clearly marked with a flag. Mice with normal vision typically swim straight toward the visual cue; those with impaired vision take more time and swim a greater distance to find the platform through random exploration of the arena.

2.5.1 Hearing

Hearing can be tested through responses to audible cues. Mice exposed to a test tone respond with specific patterns of responses including backward flicks of the external ear (pinnae—the "Preyer" reflex). Typically the preyer reflex is followed by a startle response that can range from contraction of the neck muscles to a leap or jump upward or backward. Consistency of test tone is an important control; devices such as the "Clickbox" (Institute of Hearing Research, Nottingham, UK) can provide a standardized 20-kHz stimulus tone at 90 dB [29]. More detailed assessment of hearing can be obtained through auditory-evoked brainstem response analysis (ABR), where anesthetized mice are placed in an audiometric chamber and subdermal recording electrodes placed at standardized locations over bones near the external pinna. Auditory click sound stimuli are then delivered, usually consisting of stimuli trains of varied frequency and amplitude. Evoked brainstem electrical responses are recorded through the implanted electrodes, then amplified and analyzed [29].

2.5.2 Olfaction

Olfactory responses to a social stimulus can be measured as described above using olfactory dishabituation assays. Olfaction of a nonsocial stimulus can be measured as described by Moretti et al.

[30] with a simple test where subjects are introduced individually to a novel cage with a buried piece of peanut butter cookie—a highly desirable olfactory treat. Direct observation and timing or video analysis software can be used to measure the latency to contact with the target.

2.5.3 Tests of Somatic Sensation

Tactile sensation is generally quantified through tests evaluating withdrawal responses to graded sets of monofilaments ("von Frey" hairs or "Semmes-Weinstein" filaments; Stoelting, Wood Dale, IL). The subject is placed on a wire mesh or perforated Plexiglas testing table, covered with plastic containers that restrict their movement, and allowed to acclimate for 10–20 min. Individual filaments are then applied to the plantar, lateral, and/or dorsal aspects of the hind or fore paws of unrestrained rats (location of the testing site may be dictated by other factors, such as treatments including injury of the paw or testing site; in some studies, testing sites such as the vulva or scrotum may be relevant as well). Monofilaments are applied perpendicular to the paw surface with sufficient force to cause a slight bending of the filament in increasing order of intensity until the mouse responds by withdrawal of the paw from the stimulus. Mechanical stimulation is typically repeated three times at 5–10 min intervals. Strategies for using filaments of increasing diameter or randomized application of various filaments to establish the force that evokes a withdrawal response 50% of the time have been described in detail [31–33].

Responses to noxious thermal stimuli can be assessed in several ways. Older tests include placing the mouse on a warmed thermal "hot plate," where the mouse is placed into an arena with a heated, typically metal floor plate [34]. The floor is heated to a constant temperature (generally 45–50°C) and evokes paw licking and jumping behaviors. Latency to either behavior is quantified by direct observation. Since learning rapidly changes the latency to response to the heated floor arena, subjects are usually only tested once. Similar approaches can be used to test for cold thermal sensitivity [35].

Alternatively, thermal responsiveness can be quantified using heat applied focally to the dorsal tail ("tail-flick testing") or the plantar aspect of paws (thermal analgesiometry). The refinements offered by paw-testing devices have largely displaced tail-flick testing, which requires restraint of the subject [36]. Using a popular thermal analgesiometer (UARDG; Department of Anesthesiology, University of California San Diego, La Jolla, CA) [37, 38], freely moving mice are placed under plastic chambers on an apparatus with a glass table-like top. Underneath the glass, a high-intensity light beam can then be focused on the plantar surface of a hind paw or tail as a noxious thermal stimulus. Baseline measurements are usually taken for each animal before any treatments and can be repeated at various later time points.

Both approaches are useful, but have important differences. Thermal analgesiometry measuring withdrawal threshold is measuring spinal reflex activity that often relies largely on robust monosynaptic reflexes. However, hot- or cold-plate responses require much more complex, supraspinally integrated activity. Nuances in different disruptions of responsiveness to each test can be revealing about the level of CNS function disrupted by a given gene mutation.

Models of inflammatory and neuropathic pain are numerous and widely used to define the responsiveness of mice to stimuli associated with pain. Some are very complex, such as the surgical models to establish long-lasting neuropathic hyperalgesia or allodynia in specific anatomical regions, while some are very simple, such as injection of a noxious substance into the mouse, usually subcutaneously in a single hind paw. Inflammation can be induced with any one of numerous substances, but a well-defined and widely used model of inflammatory pain consists of injection of a small volume (about 25 μl) of dilute formalin into the hind paw of the mouse. An advantage of this test is that it evokes a pattern of spontaneous pain-related behaviors such as paw licking and biting and hind limb flinches with biphasic peaks that are differentially responsive to centrally acting versus anti-inflammatory analgesic drugs. In this test, nociceptive behaviors are usually monitored for 60 min after hind paw treatment. The mouse is placed in an observation chamber and hind paw flinches, as described by Wheeler-Aceto et al. [39], counted by direct observation for 1 min period by an observer blind to drug and hind paw treatments. Typically, data on spontaneous pain-related behaviors will be collected in three periods: 0–10 min, 10–30 min, and 30–60 min after injection of formalin into the hind paw. This corresponds to the peak period of responsiveness during the early (first) and late (second) phase of the formalin test as well as the interphase, when modulatory activity produces a transient lull in pain-related behaviors between the early and late phases [32, 40].

2.6 Tests of Locomotor Function

2.6.1 Open Field Locomotion

The most straightforward and simplest tests of general locomotor activity are based on observation of the subject mouse's movements when placed in a novel open arena. The mouse is introduced into the center of the arena, and typically will move rapidly to a wall of the enclosure before exploring the entire arena, usually remaining close to the walls. This thigmotactic (wall-hugging) behavior is a function of the anxiety of the mouse toward open exposure to potential predators. As the mouse becomes more familiar with the new environment, it increasingly ventures intermittently into the center of the arena. The mouse has a natural drive to explore its new surroundings, and this inherent behavior can be leveraged to assess the patterns of movement the mouse exhibits, such as total distance traveled, rearing, circling, bouts of low motility, and stereotypic behaviors (e.g., grooming). The open field arenas for mice can be circular or square, and are typically divided into areas to facilitate observation of the patterns of movement around the arena. While this approach facilitates easy observation and quantification of basic locomotor activities, open field testing can be confounded by features such as intensity of lighting or residual olfactory cues from previous test subjects.

Automated recording and measurement of locomotor activity provides excellent advantages over direct observation and recording by the investigator. Devices are available that automatically record mouse movement through breakage of photocell beams, electromagnetic detection, or video-tracking software [41–43]. Some of these technologies can be adapted to home cages instead of novel test arenas; this avoids anxiety-evoked thigmotaxis, but loses the advantage of evoking locomotion by exposure to a novel environment. A distinct advantage of automated systems includes labor-reducing standardization of objective, unbiased quantification and recording of locomotion, and multiple options regarding data analysis.

2.6.2 Force-Plate Actimetry

For much more detailed analysis of locomotor activity, exploratory behaviors, stereotypy, tremor, and ataxia, the BASi Force-Plate Actimeter (BASi, Mount Vernon, IN) provides an incredibly rich data stream that records and quantifies fine details of movement. The actimeter arena consists of a square, stiff horizontal plate attached at the corners to force transducers that record the center of mass of the animal subject at 100 Hz. A Plexiglas enclosure rests a few millimeters above the plate to create a transparent enclosure. Animals are placed into the testing arena and allowed to move freely for the testing period (usually 10 min for initial screening purposes). When the test subject moves on the plate, its movements are sensed by the transducers, whose signals are processed by a computer to detect and score a wide range of behaviors or behavioral attributes, including behavioral events occurring at user-defined frequencies or amplitudes. These include, but are not limited to, stereotypies, tremor, startle, or ataxia. It simultaneously records larger movements such as locomotion (with mm scale resolution) and fine movements (with frequency resolution up to 100 Hz). Analysis of the data gathered during this time reveals several parameters such as total distance traveled, area traveled, bouts of low mobility, focused stereotypies, left and right turns, and other aspects of both general locomotion and small body movements, up to and including whisking, respiration, and even heartbeat [44–48]. This apparatus has been particularly useful in analyzing rotational behaviors, inherent or drug-evoked tremor, and seizure-like activity in rodents [49].

2.6.3 Balance and Coordination

Ataxia, or loss of the mouse's ability to maintain its balance and to coordinate complex movement during an evoked locomotor challenge, can be assessed using an apparatus known as the Rotarod. The Rotarod was designed to automatically assess neurological deficits in rodents [50] and is one of the most commonly used tests of motor function in mice. A typical example of this rotating treadmill (Accuscan, Columbus, OH) consists of a suspended drum that rotates at accelerating speed of 4–40 rpm, which are controlled by a computer controller. The rotating rod is elevated several inches above the floor of the testing arena, which leverages the mouse's natural fear of falling as a motivation to maintain its balance on the rod by locomoting at the appropriate speed. In testing for ataxia, mice are typically exposed to the Rotarod for several [3–5] training exposures of approximately 5 min each, with intervals between training sessions of 10–15 min. The mouse's ability to balance on the rod is then assessed during a subsequent test trial on the Rotarod. Data are gathered as latency (sec) to fall from the rod, which may be constrained by a preset maximum time (e.g., 3 min).

The rotation speed and acceleration of the rod varies widely between different studies, and specific speeds or durations of testing may be optimal for demonstrating specific types of impairment. Tests on an accelerating rod force locomotor challenges more rapidly, but fixed speed testing can be important in assessing performance across multiple specific rotation speeds or for assessing fatigue at longer test durations [51]. Rotarod testing has shown that transgenic phenotypes emerge as a function of age, task difficulty, and sensitivity of the test to ataxia [52]. Importantly, Rotarod testing can be confounded by several factors including body mass and size, fatigue, learning, subjects' clinging to the rod and "riding" instead of locomoting, or simple refusal of the individual subject to perform the test.

Ataxia and balance can also be tested using a simpler "beam walk" analysis. This test was developed to assess aging associated motor deficits [53], but has been useful in assessing balance and coordination in lesioned or mutant mice [52, 54]. Test subjects are assessed for disturbances in normal gait and balance by observing their ability to traverse a 75-cm long, 15 mm diameter wooden dowel elevated 30 cm and ending in a dark goal box [55]. Mice are trained to traverse the beam and enter the dark enclosure located at the end of the beam, which can be baited with a desirable treat. Training typically requires 5–10 exposures to the walking task initiated at increasing distances from the dark box. After training, animals are videotaped traversing the beam three times. The number of foot slip errors for each trial are counted and summed. The difficulty of the beam walk task can be modified by using beams of different cross-sectional diameter or shape to vary the sensitivity of the test.

2.6.4 Grip Strength

Fore limb, hind limb, and whole-body grip strength are best measured using a force transducer-based apparatus (e.g., Animal Grip Strength System, San Diego Instruments, San Diego, CA). The grip strength system utilizes an open wire grid that the suspended mouse grabs with available paws. The amount of resistance offered by the animal when pulled from the wire grid (grip strength) is measured by force transducers attached to the wire grid. Grip strength is recorded as the maximum amount of force that the mouse exerts while holding on to the grid and is measured in triplicate trials and normalized to body weight where appropriate. Grip strength testing has revealed progressive loss of grip strength in several mutant mouse models [56, 57].

Alternative tests use approaches to test the mouse's ability to cling to an inverted or tilted wire grid or hang from a wire with its forepaws until its grip fails [58]. These older approaches, however, include aspects of locomotion and motivation that affect the test in ways much more complex than simple paw grip strength and are often confounded by locomotion across the grid/wire or willingness of the mouse to perform the task.

2.6.5 Gait and Posture

Other detailed aspects of motor coordination and synchrony can be gained by observing the mouse's gait during walking. State-of-the-art approaches use video recordings of the mouse running on a clear treadmill or runway [58]. An older, but still commonly used assay of gait is the "footprint" test, where the fore and hind paws of the mouse are dipped in ink before ambulating freely on a paper arena floor [52, 59]. The resulting footfall patterns can be analyzed for measurements including stride length, base width, fore/hind overlap, and splay. Data gathering for footprint analysis by inking paws is both imprecise and labor intensive and does not restrict locomotion to specific speeds. While informative, this approach does not reveal the temporal aspects of gait dynamics that are captured using video assessment of movement across strides over time.

Perhaps the most well-characterized instrument for evaluating functional deficits in gait is the Basso, Beattie, and Bresnahan (BBB) scale, developed for assessing open field locomotion in rats with spinal injury [60]. This gait assessment tool has been specifically adapted for studying mouse behavior through the creation of the 10-point assessment Basso Mouse Scale (BMS) [61]. The BMS scale has been validated, is widely used, and is highly standardized, but only addresses a subset of readily observable attributes of gait. BMS scoring includes observation of basic movements limited to ankle, knee, and hip range of motion and plantar placement of the hind limb with or without weight support. Additionally, observation of coordination, hind paw position during locomotion, toe clearance, tail position, and trunk stability are scored.

Automated gait analysis systems can reveal aspects of gait that may be too rapid, complex, or subtle for direct visual observation. Furthermore, automated measurement of gait dynamics removes the observer as a source of variability. Good examples of useful automated gait analysis systems include the CatWalk™ [62–65] or DigiGait™ (Mouse Specifics, Inc.) or TreadScan [66, 67] systems, which consist of a camera mounted beneath a variable speed transparent treadmill or runway. These systems allow automated quantitation of numerous detailed attributes of gait; importantly, with some systems, data can be gathered at specific user-selected treadmill speeds.

2.7 Tests of Cognitive Function

Tests are available that assess the cognitive capacity and function of rodent subjects in varying degrees of complexity, from simple maze performance to complex operant training/conditioning paradigms. The simplest of these tests can be easily incorporated into screening protocols, but may only reveal severe deficits in cognitive functions such as working memory or anxiety-related phenotypes.

2.7.1 Working Memory

Testing locomotor patterns in a simple, three-arm Y-maze can be utilized to measure working and reference memory. Animals are placed into the center zone of a Y-shaped arena and allowed to freely explore for 5 min. Entries into the maze arms and percent alternation between the arms are recorded. Wild-type mice explore the entire maze, and typically alternate their exploration of the separate arms of the maze; failure of the mutant mouse to do either can reveal compromised working or reference memory function.

Similarly, T-mazes or more complex multiple T-mazes can test the acquisition of memory and memory retention in mice. A video capture system is used to observe the mouse navigating freely through the maze. External spatial cues can be used to alter the difficulty of the task. The time required for the mouse to reach the goal box or food reward at the end of the maze is recorded. With initial exposure to the maze, repeated multiple trials through the maze are used to observe memory formation. Memory retention is tested after a prolonged temporal break in testing trials.

2.7.2 Memory Acquisition and Retention

The Morris water navigation task, also known as the Morris water maze (MWM), is a behavioral procedure used to study spatial learning and memory [68, 69]. The resources needed to conduct MWM testing are minimal, but the task is complex and flexible enough to provide in-depth assessments of memory acquisition, retention, and decay, and requires spatial working memory that is disrupted by aberrant development or damage to cortical or hippocampal regions of the brain [70, 71]. Morris water maze testing has been shown to be useful in quantitating cognitive deficits in mutant mouse models [72].

In basic MWM testing, the subject is placed in a large circular pool of water and swims to find a visible or invisible platform just below the water's surface that allows it to escape the water by climbing onto the platform. This task is well suited to use for testing mice since they are not natural swimmers like rats, but dislike immersion in water of any temperature; this provides inherent motivation to perform the task without external motivating stimuli such as electrical shock or food deprivation. The mobility of the platform facilitates assays of learning, relearning, memory retention, and spatial navigation. Specialized organization of the initial exposures of the mouse to the MWM in training trials and the timing of probe trials that test the retention of memory of the location of the hidden platform provides paradigms for quantifying different cognitive functions. For example, a simple test of vision, as described above, is to expose the mouse to a swim trial with a clearly visible platform marked with a flag. Cognitive flexibility can be assessed using a water maze paradigm in which the hidden platform is continually relocated.

In the simplest assays of learning and memory retention, mice are trained to find the hidden platform in four learning trials over the course of 30 min per day for 4 days. Memory acquisition is observed through shorter latencies to reach the goal platform and decreased distance traveled to the platform. Data can be gathered using direct observation and timing, but much more detailed analysis of the mouse's behavior can be measured using analysis of video recordings of trials captured by a camera mounted above the swim arena. Memory retention can then be observed through repeated single trials administered on later days. Working memory function can be assessed by comparing time spent in goal versus nongoal quadrants of the MWM.

2.7.3 Conditioned Fear Responses

The ability of the test subjects to learn and remember the association of a novel environment or an acoustic cue with an unpleasant foot shock can be measured using an arena with a shock-grid floor that can deliver a noxious foot shock with or without an audible test tone [73]. This contextual fear conditioning is achieved by allowing a test subject mouse to explore a novel, open arena with a shock floor. After 2 min of exploration, a single foot shock (e.g., 0.35 mA for 1 s) is then delivered. The mouse is removed and returned to its home cage. At various relevant time points (e.g., 1 h or 1 day) following training, the mouse is returned to the contextual arena for 4 min. Locomotion and freezing behaviors are scored in 5 s intervals through direct observation or automated recording, often using infrared beam breaks in the arena. Typically, wild-type mice quickly associate the novel environment and/or test tone with an anticipated foot shock, and respond with freezing behaviors where locomotion is stopped entirely. Impairments in memory can be assessed by the relative inability of test subjects to gain or retain freezing behaviors associated with the test stimulus.

2.7.4 Prepulse Inhibition (PPI) Testing

Mice will exhibit a reflexive startle response to unanticipated, sudden, loud auditory stimuli. However, the startle response can be inhibited if the stimulus is preceded by a warning stimulus ("prepulse"). The ability of prepulse stimuli to inhibit startle responses is modified in several neurological and psychiatric conditions such as autism and schizophrenia. In these disorders, the acoustic startle response is less completely inhibited by prepulse stimuli, suggesting perturbations of higher order central cognitive processes that result in sensory gating of the acoustic startle motor reflex [74–76].

Measurements of the startle reflex are usually performed using a microprocessor-controlled auditory stimulus chambers where the somewhat confining animal chamber has a force transducer-mounted floor that measures the magnitude of jump-like startle reactions to brief, intense acoustic tones (usually 90–120 dB bursts of white noise lasting less than 100 ms).

Calibrated prepulse tones are delivered in a stairstep, pseudorandomized order [77]. To assess inhibition by the brief warning prepulse stimulus, a short, low-intensity burst of white noise is delivered 20–500 ms preceding the startle stimulus. Given multiple exposures, the negative association between the white noise prepulse and evoked startle responses should strengthen and the animal should respond less robustly to the startle.

Testing of PPI is subject to confound by several factors, including strain and sex differences, body weight, and aging. It is important to be aware that some mouse strains develop deafness that can robustly affect both the primary and the prepulse startle responses [78]. Nonetheless, automated assessment of acoustic startle and PPI is used widely for screening the phenotype of novel mutant mouse lines and has demonstrated usefulness in identifying subtle aspects of the phenotype of mutant mouse models of neurological disease [52, 79].

2.8 Tests of Anxiety and Emotional Reactivity

Many well-validated behavioral tests exist for measuring the anxiety or emotion-like reactivity of mice to stimuli that are stressful or leverage some ethological conflict between various patterned behaviors in the mouse [80]. These tests can be used to emulate both the positive and negative symptoms of human mood disorders in mouse models.

2.8.1 Place Preference

The anxiety of mice toward light can be measured by quantifying the frequency of emergence from the enclosed, darkened side of an arena into the open, more brightly lit side through a small opening between them. This light/dark place preference test takes advantage of the animals' natural photophobia to measure light aversion, which can be heightened in stressed animals or those with an anxiety-related phenotype. We currently use modified force-plate actimeters for this purpose [81]. The intensity of lighting in the "light side" of the place-preference arena can be a very important aspect of this test; light levels may need to be relatively low (~150 lux) in order for subjects to explore both sides equally; more intense light levels will result in more time spent in the darkened chamber and may produce a ceiling effect where animals avoid the light despite anxiolytic treatments or phenotypes.

2.8.2 Thigmotaxis

During the assessment of locomotor activity in exploration of open field arenas, patterns of locomotion associated with increased wall-hugging (thigmotaxis) and suppressed rearing behaviors can be quantified as anxiety-related behaviors, as described above. These behavioral endpoints are common features extracted from data gathered and analyzed by automated locomotor testing arenas. In this context, arena size and lighting are of primary importance as the mouse's preference to avoid the center of the arena largely reflects an aversion to light. Specific assay conditions optimized for measurement of anxiety-related thigmotaxis have been described [82, 83].

2.8.3 Elevated Plus Maze

A similar test uses an elevated, plus-shaped (+) maze apparatus with two open and two wall-enclosed arms [84]. The mouse is introduced into the neutral center of the maze and then the patterns of movement into the various arms recorded by direct observation or automated video tracking software. The test relies on the general aversion of mice to both ledges and open spaces. In navigating the elevated plus maze, anxiety-related behaviors are expressed by the larger proportion of time spent in the closed maze arms (time in closed arms/total time) and less frequent entries into the open arms (entries into open arms/total arm entries). The total number of arm entries can also be used as proxy measurement of general locomotor activity [85]. Elevated plus maze testing can be confounded in strains of mice with pronounced circling behaviors or across groups of mice with significant differences in total locomotor output. The elevated plus maze has been used to successfully characterize strain differences in mice [86].

2.9 Data Collection and Reporting Strategies

Ultimately, the usefulness and validity of behavioral data collected using these approaches depends on more than just the local quality control of experimental design, experimenter/observer training and consistency, and sound apparatus and automated data collection systems. Perhaps the largest single challenge facing behavioral researchers currently is comparison of results from experiments conducted at different institutions or testing sites. Automated data acquisition and analysis obviates much of the variability introduced by individual observers, but considerable differences in the outcomes of behavioral tests are nonetheless apparent when results from different investigator teams are compared. In the field of complex behavioral experimentation, significant variability can be introduced by "subtle" differences between institutions resulting from varied season, diet, housing conditions, room temperatures, etc. Small differences in animal handling techniques, animal transport, time of day, order of testing, and attributes of the investigator can contribute to significant variability in experimental results. Accordingly, increasing attention and effort is being paid toward establishing standardized testing conditions and paradigms for implementation across multiple testing sites phenotyping experimental strains of mice. Guidelines are emerging to attempt to increase reproducibility in behavioral studies and improve the communication between researchers and sharing of research findings. For example, the Animal Research: Reporting of In Vivo Experiments (ARRIVE) guidelines are increasingly used to enhance transparency in methods and data reporting and have become a key feature of the standardized reporting approaches used by publishers and managers of shared databases [87]. Consortia have been established to try and create large-scale, multicenter phenotyping pipelines that will promote reproducible data collection and effective data dissemination. Examples such as the EUMORPHIA and EUMODIC consortia that use EMPReSSslim phenotyping paradigms reflect state-of-the-art efforts in this regard [88–90]. Large-scale efforts creating and characterizing mutant mice such as the International Knockout Mouse Consortium [91, 92] and the International Mouse Phenotyping Consortium [93] will be dependent on phenotyping centers with the necessary expertise, resources, and capacity to provide the necessary phenotyping. In conclusion, behavioral phenotyping of mice with novel targeted gene mutations is a key initial step in determining the potential impact of novel mouse models and their relevance to the development of new therapies for human disease. Many of the tests described here can be used by investigators in their own facilities to begin the behavioral screening of the mutant mice they develop, and most techniques can be established with some guiding input from an investigator with adequate knowledge and background. For complex behavioral tests or extensive phenotyping goals, collaboration with an established rodent behavioral core facility or investigators currently using the specific phenotyping approaches can be the most efficient strategy.

Acknowledgments

The author thanks Michelle K. Winter for her expert behavioral core management and editorial assistance and the Rodent Behavior Facility of the Kansas Intellectual and Developmental Disabilities Research Center (HD090216) for the long-term funding.

References

1. Mogil JS, Sternberg WF, Kest B, Marek P, Liebeskind JC (1993) Sex differences in the antagonism of swim stress-induced analgesia: effects of gonadectomy and estrogen replacement. Pain 53(1):17–25
2. Greenspan JD, Craft RM, LeResche L, Arendt-Nielsen L, Berkley KJ, Fillingim RB, Gold MS, Holdcroft A, Lautenbacher S, Mayer EA, Mogil JS, Murphy AZ, Traub RJ, Consensus Working Group of the Sex G, Pain SIGotI (2007) Studying sex and gender differences in pain and analgesia: a consensus report. Pain 132(Suppl 1):S26–S45. https://doi.org/10.1016/j.pain.2007.10.014
3. Meziane H, Ouagazzal AM, Aubert L, Wietrzych M, Krezel W (2007) Estrous cycle effects on behavior of C57BL/6J and BALB/cByJ female mice: implications for phenotyping strategies. Genes Brain Behav 6(2):192–200. https://doi.org/10.1111/j.1601-183X.2006.00249.x
4. Mogil JS, Chanda ML (2005) The case for the inclusion of female subjects in basic science studies of pain. Pain 117(1–2):1–5. https://doi.org/10.1016/j.pain.2005.06.020
5. Crawley JN (1999) Behavioral phenotyping of transgenic and knockout mice: experimental design and evaluation of general health, sensory functions, motor abilities, and specific behavioral tests. Brain Res 835(1):18–26
6. Castelhano-Carlos MJ, Sousa N, Ohl F, Baumans V (2010) Identification methods in newborn C57BL/6 mice: a developmental and behavioural evaluation. Lab Anim 44 (2):88–103. https://doi.org/10.1258/la.2009.009044
7. Crawley JN (2003) Behavioral phenotyping of rodents. Comp Med 53(2):140–146
8. Crawley JN (2008) Behavioral phenotyping strategies for mutant mice. Neuron 57 (6):809–818. https://doi.org/10.1016/j.neuron.2008.03.001
9. Crawley JN, Paylor R (1997) A proposed test battery and constellations of specific behavioral paradigms to investigate the behavioral phenotypes of transgenic and knockout mice. Horm Behav 31(3):197–211. https://doi.org/10.1006/hbeh.1997.1382
10. McIlwain KL, Merriweather MY, Yuva-Paylor LA, Paylor R (2001) The use of behavioral test batteries: effects of training history. Physiol Behav 73(5):705–717
11. Brooks SP, Pask T, Jones L, Dunnett SB (2004) Behavioural profiles of inbred mouse strains used as transgenic backgrounds. I: motor tests. Genes Brain Behav 3 (4):206–215. https://doi.org/10.1111/j.1601-183X.2004.00072.x
12. Rogers DC, Fisher EM, Brown SD, Peters J, Hunter AJ, Martin JE (1997) Behavioral and functional analysis of mouse phenotype: SHIRPA, a proposed protocol for comprehensive phenotype assessment. Mamm Genome 8 (10):711–713
13. Irwin S (1968) Comprehensive observational assessment: Ia. A systematic, quantitative procedure for assessing the behavioral and physiologic state of the mouse. Psychopharmacologia 13(3):222–257
14. Rogers DC, Peters J, Martin JE, Ball S, Nicholson SJ, Witherden AS, Hafezparast M, Latcham J, Robinson TL, Quilter CA, Fisher EM (2001) SHIRPA, a protocol for behavioral assessment: validation for longitudinal study of neurological dysfunction in mice. Neurosci Lett 306(1–2):89–92
15. Lalonde R, Strazielle C (2001) Motor performance and regional brain metabolism of spontaneous murine mutations with cerebellar atrophy. Behav Brain Res 125(1–2):103–108
16. Stasi K, Mitsacos A, Triarhou LC, Kouvelas ED (1997) Cerebellar grafts partially reverse amino acid receptor changes observed in the cerebellum of mice with hereditary ataxia: quantitative autoradiographic studies. Cell Transplant 6 (3):347–359
17. Rogers DC, Jones DN, Nelson PR, Jones CM, Quilter CA, Robinson TL, Hagan JJ (1999) Use of SHIRPA and discriminant analysis to characterise marked differences in the behavioural phenotype of six inbred mouse strains. Behav Brain Res 105(2):207–217
18. Masuya H, Inoue M, Wada Y, Shimizu A, Nagano J, Kawai A, Inoue A, Kagami T, Hirayama T, Yamaga A, Kaneda H, Kobayashi K, Minowa O, Miura I, Gondo Y, Noda T, Wakana S, Shiroishi T (2005) Implementation of the modified-SHIRPA protocol for screening of dominant phenotypes in a large-scale ENU mutagenesis program. Mamm Genome 16(11):829–837. https://doi.org/10.1007/s00335-005-2430-8
19. Oellrich A, Meehan TF, Parkinson H, Sarntivijai S, White JK, Karp NA (2016) Reporting phenotypes in mouse models when considering body size as a potential confounder. J Biomed Semantics 7:2. https://doi.org/10.1186/s13326-016-0050-8
20. Wills GD, Wesley AL, Moore FR, Sisemore DA (1983) Social interactions among rodent

conspecifics: a review of experimental paradigms. Neurosci Biobehav Rev 7(3):315–323
21. Blanchard DC, Blanchard RJ (2003) What can animal aggression research tell us about human aggression? Horm Behav 44(3):171–177
22. Nadler JJ, Moy SS, Dold G, Trang D, Simmons N, Perez A, Young NB, Barbaro RP, Piven J, Magnuson TR, Crawley JN (2004) Automated apparatus for quantitation of social approach behaviors in mice. Genes Brain Behav 3(5):303–314. https://doi.org/10.1111/j.1601-183X.2004.00071.x
23. Moy SS, Nadler JJ, Magnuson TR, Crawley JN (2006) Mouse models of autism spectrum disorders: the challenge for behavioral genetics. Am J Med Genet C Semin Med Genet 142C (1):40–51. https://doi.org/10.1002/ajmg.c.30081
24. Blundell J, Blaiss CA, Etherton MR, Espinosa F, Tabuchi K, Walz C, Bolliger MF, Sudhof TC, Powell CM (2010) Neuroligin-1 deletion results in impaired spatial memory and increased repetitive behavior. J Neurosci 30 (6):2115–2129. https://doi.org/10.1523/JNEUROSCI.4517-09.2010
25. Page DT, Kuti OJ, Sur M (2009) Computerized assessment of social approach behavior in mouse. Front Behav Neurosci 3:48. https://doi.org/10.3389/neuro.08.048.2009
26. Cabib S, Pascucci T, Ventura R, Romano V, Puglisi-Allegra S (2003) The behavioral profile of severe mental retardation in a genetic mouse model of phenylketonuria. Behav Genet 33 (3):301–310
27. Brigman JL, Graybeal C, Holmes A (2010) Predictably irrational: assaying cognitive inflexibility in mouse models of schizophrenia. Front Neurosci 4:13. https://doi.org/10.3389/neuro.01.013.2010
28. Silverman JL, Yang M, Lord C, Crawley JN (2010) Behavioural phenotyping assays for mouse models of autism. Nat Rev Neurosci 11(7):490–502. https://doi.org/10.1038/nrn2851
29. Hardisty-Hughes RE, Parker A, Brown SD (2010) A hearing and vestibular phenotyping pipeline to identify mouse mutants with hearing impairment. Nat Protoc 5 (1):177–190. https://doi.org/10.1038/nprot.2009.204
30. Moretti P, Bouwknecht JA, Teague R, Paylor R, Zoghbi HY (2005) Abnormalities of social interactions and home-cage behavior in a mouse model of Rett syndrome. Hum Mol Genet 14(2):205–220. https://doi.org/10.1093/hmg/ddi016
31. Brennan TJ, Vandermeulen EP, Gebhart GF (1996) Characterization of a rat model of incisional pain. Pain 64(3):493–501
32. Winter MK, McCarson KE (2005) G-protein activation by neurokinin-1 receptors is dynamically regulated during persistent nociception. J Pharmacol Exp Ther 315(1):214–221. https://doi.org/10.1124/jpet.105.089565
33. McCarson KE, Ralya A, Reisman SA, Enna SJ (2005) Amitriptyline prevents thermal hyperalgesia and modifications in rat spinal cord GABA(B) receptor expression and function in an animal model of neuropathic pain. Biochem Pharmacol 71(1–2):196–202. https://doi.org/10.1016/j.bcp.2005.10.026
34. Eddy NB, Leimbach D (1953) Synthetic analgesics. II. Dithienylbutenyl- and dithienylbutylamines. J Pharm Exp Therap 107 (3):385–393
35. Yalcin I, Charlet A, Freund-Mercier MJ, Barrot M, Poisbeau P (2009) Differentiating thermal allodynia and hyperalgesia using dynamic hot and cold plate in rodents. J Pain 10(7):767–773. https://doi.org/10.1016/j.jpain.2009.01.325
36. Dewey WL, Snyder JW, Harris LS, Howes JF (1969) The effect of narcotics and narcotic antagonists on the tail-flick response in spinal mice. J Pharm Pharmacol 21(8):548–550
37. Dirig DM, Salami A, Rathbun ML, Ozaki GT, Yaksh TL (1997) Characterization of variables defining hindpaw withdrawal latency evoked by radiant thermal stimuli. J Neurosci Methods 76 (2):183–191
38. Hargreaves K, Dubner R, Brown F, Flores C, Joris J (1988) A new and sensitive method for measuring thermal nociception in cutaneous hyperalgesia. Pain 32(1):77–88
39. Wheeler-Aceto H, Porreca F, Cowan A (1990) The rat paw formalin test: comparison of noxious agents. Pain 40:229–238
40. Dubuisson D, Dennis SG (1977) The formalin test: a quantitative study of the analgesic effects of morphine, meperidine, and brain stem stimulation in rats and cats. Pain 4(2):161–174
41. Grusser C, Grusser-Cornehls U (1998) Improvement in motor performance of Weaver mutant mice following lesions of the cerebellum. Behav Brain Res 97(1–2):189–194
42. Thifault S, Lalonde R, Sanon N, Hamet P (2001) Longitudinal analysis of motor activity and coordination, anxiety, and spatial learning in mice with altered blood pressure. Brain Res 910(1–2):99–105
43. Drai D, Kafkafi N, Benjamini Y, Elmer G, Golani I (2001) Rats and mice share common

ethologically relevant parameters of exploratory behavior. Behav Brain Res 125 (1–2):133–140

44. Crosland KA, Zarcone JR, Schroeder S, Zarcone T, Fowler S (2005) Use of an antecedent analysis and a force sensitive platform to compare stereotyped movements and motor tics. Am J Mental Retard 110(3):181–192. https://doi.org/10.1352/0895-8017(2005)110<181:UOAAAA>2.0.CO;2
45. Fowler SC, Birkestrand B, Chen R, Vorontsova E, Zarcone T (2003) Behavioral sensitization to amphetamine in rats: changes in the rhythm of head movements during focused stereotypies. Psychopharmacology 170(2):167–177. https://doi.org/10.1007/s00213-003-1528-5
46. Fowler SC, Zarcone TJ, Levant B (2010) Methylphenidate attenuates rats' preference for a novel spatial stimulus introduced into a familiar environment: assessment using a force-plate actometer. J Neurosci Methods 189 (1):36–43. https://doi.org/10.1016/j.jneumeth.2010.03.014
47. McKerchar TL, Zarcone TJ, Fowler SC (2006) Use of a force-plate actometer for detecting and quantifying vertical leaping induced by amphetamine in BALB/cJ mice, but not in C57BL/6J, DBA/2J, 129X1/SvJ, C3H/HeJ, and CD-1 mice. J Neurosci Methods 153(1):48–54. https://doi.org/10.1016/j.jneumeth.2005.10.002
48. Smittkamp SE, Brown JW, Stanford JA (2008) Time-course and characterization of orolingual motor deficits in B6SJL-Tg(SOD1-G93A)1Gur/J mice. Neuroscience 151(2):613–621. https://doi.org/10.1016/j.neuroscience.2007.10.017
49. Fowler SC, Miller BR, Gaither TW, Johnson MA, Rebec GV (2009) Force-plate quantification of progressive behavioral deficits in the R6/2 mouse model of Huntington's disease. Behav Brain Res 202(1):130–137. https://doi.org/10.1016/j.bbr.2009.03.022
50. Dunham NW, Miya TS (1957) A note on a simple apparatus for detecting neurological deficit in rats and mice. J Am Pharm Assoc Am Pharm Assoc 46(3):208–209
51. Monville C, Torres EM, Dunnett SB (2006) Comparison of incremental and accelerating protocols of the rotarod test for the assessment of motor deficits in the 6-OHDA model. J Neurosci Methods 158(2):219–223. https://doi.org/10.1016/j.jneumeth.2006.06.001
52. Carter RJ, Lione LA, Humby T, Mangiarini L, Mahal A, Bates GP, Dunnett SB, Morton AJ (1999) Characterization of progressive motor deficits in mice transgenic for the human Huntington's disease mutation. J Neurosci 19 (8):3248–3257
53. Wallace JE, Krauter EE, Campbell BA (1980) Motor and reflexive behavior in the aging rat. J Gerontol 35(3):364–370
54. Kiernan BW, Garcion E, Ferguson J, Frost EE, Torres EM, Dunnett SB, Saga Y, Aizawa S, Faissner A, Kaur R, Franklin RJ, ffrench-Constant C (1999) Myelination and behaviour of tenascin-C null transgenic mice. Eur J Neurosci 11(9):3082–3092
55. Russell KL, Kutchko KM, Fowler SC, Berman NE, Levant B (2011) Sensorimotor behavioral tests for use in a juvenile rat model of traumatic brain injury: assessment of sex differences. J Neurosci Methods 199(2):214–222. https://doi.org/10.1016/j.jneumeth.2011.05.008
56. Glynn D, Drew CJ, Reim K, Brose N, Morton AJ (2005) Profound ataxia in complexin I knockout mice masks a complex phenotype that includes exploratory and habituation deficits. Hum Mol Genet 14(16):2369–2385. https://doi.org/10.1093/hmg/ddi239
57. Lorivel T, Hilber P (2007) Motor effects of delta 9 THC in cerebellar Lurcher mutant mice. Behav Brain Res 181(2):248–253. https://doi.org/10.1016/j.bbr.2007.04.011
58. Heglund NC, Taylor CR, McMahon TA (1974) Scaling stride frequency and gait to animal size: mice to horses. Science 186 (4169):1112–1113
59. Clarke KA, Still J (2001) Development and consistency of gait in the mouse. Physiol Behav 73(1–2):159–164
60. Basso DM, Beattie MS, Bresnahan JC (1995) A sensitive and reliable locomotor rating scale for open field testing in rats. J Neurotrauma 12 (1):1–21
61. Basso DM, Fisher LC, Anderson AJ, Jakeman LB, McTigue DM, Popovich PG (2006) Basso Mouse Scale for locomotion detects differences in recovery after spinal cord injury in five common mouse strains. J Neurotrauma 23 (5):635–659. https://doi.org/10.1089/neu.2006.23.635
62. Hamers FP, Lankhorst AJ, van Laar TJ, Veldhuis WB, Gispen WH (2001) Automated quantitative gait analysis during overground locomotion in the rat: its application to spinal cord contusion and transection injuries. J Neurotrauma 18(2):187–201. https://doi.org/10.1089/08977150150502613
63. Koopmans GC, Deumens R, Honig WM, Hamers FP, Steinbusch HW, Joosten EA (2005) The assessment of locomotor function in spinal cord injured rats: the importance of objective analysis of coordination. J

Neurotrauma 22(2):214–225. https://doi.org/10.1089/neu.2005.22.214
64. Hamers FP, Koopmans GC, Joosten EA (2006) CatWalk-assisted gait analysis in the assessment of spinal cord injury. J Neurotrauma 23 (3–4):537–548. https://doi.org/10.1089/neu.2006.23.537
65. Jeong MA, Plunet W, Streijger F, Lee JH, Plemel JR, Park S, Lam CK, Liu J, Tetzlaff W (2011) Intermittent fasting improves functional recovery after rat thoracic contusion spinal cord injury. J Neurotrauma 28 (3):479–492. https://doi.org/10.1089/neu.2010.1609
66. Leblond H, L'Esperance M, Orsal D, Rossignol S (2003) Treadmill locomotion in the intact and spinal mouse. J Neurosci 23 (36):11411–11419
67. Myers SA, DeVries WH, Gruenthal MJ, Andres KR, Hagg T, Whittemore SR (2012) Sildenafil improves epicenter vascular perfusion but not hindlimb functional recovery after contusive spinal cord injury in mice. J Neurotrauma 29 (3):528–538. https://doi.org/10.1089/neu.2011.2036
68. Morris R (1984) Developments of a water-maze procedure for studying spatial learning in the rat. J Neurosci Methods 11(1):47–60
69. Hodges H (1996) Maze procedures: the radial-arm and water maze compared. Brain Res Cogn Brain Res 3(3–4):167–181
70. D'Hooge R, De Deyn PP (2001) Applications of the Morris water maze in the study of learning and memory. Brain Res Rev 36 (1):60–90
71. Morris RG, Garrud P, Rawlins JN, O'Keefe J (1982) Place navigation impaired in rats with hippocampal lesions. Nature 297 (5868):681–683
72. Morgan D (2009) Frontiers in neuroscience water maze tasks in mice: special reference to Alzheimer's transgenic mice. In: Buccafusco JJ (ed) Methods of behavior analysis in neuroscience. CRC Press/Taylor & Francis, Taylor & Francis Group, LLC, Boca Raton (FL)
73. Valentinuzzi VS, Kolker DE, Vitaterna MH, Shimomura K, Whiteley A, Low-Zeddies S, Turek FW, Ferrari EA, Paylor R, Takahashi JS (1998) Automated measurement of mouse freezing behavior and its use for quantitative trait locus analysis of contextual fear conditioning in (BALB/cJ x C57BL/6J)F2 mice. Learn Mem 5(4–5):391–403
74. Grillon C, Ameli R, Charney DS, Krystal J, Braff D (1992) Startle gating deficits occur across prepulse intensities in schizophrenic patients. Biol Psychiatry 32(10):939–943
75. Perriol MP, Dujardin K, Derambure P, Marcq A, Bourriez JL, Laureau E, Pasquier F, Defebvre L, Destee A (2005) Disturbance of sensory filtering in dementia with Lewy bodies: comparison with Parkinson's disease dementia and Alzheimer's disease. J Neurol Neurosurg Psychiatry 76(1):106–108. https://doi.org/10.1136/jnnp.2003.035022
76. Swerdlow NR, Paulsen J, Braff DL, Butters N, Geyer MA, Swenson MR (1995) Impaired prepulse inhibition of acoustic and tactile startle response in patients with Huntington's disease. J Neurol Neurosurg Psychiatry 58(2):192–200
77. Swerdlow NR, Geyer MA (1998) Using an animal model of deficient sensorimotor gating to study the pathophysiology and new treatments of schizophrenia. Schizophr Bull 24 (2):285–301
78. Willott JF, Turner JG, Carlson S, Ding D, Seegers Bross L, Falls WA (1998) The BALB/c mouse as an animal model for progressive sensorineural hearing loss. Hear Res 115 (1–2):162–174
79. Van Raamsdonk JM, Pearson J, Slow EJ, Hossain SM, Leavitt BR, Hayden MR (2005) Cognitive dysfunction precedes neuropathology and motor abnormalities in the YAC128 mouse model of Huntington's disease. J Neurosci 25(16):4169–4180. https://doi.org/10.1523/jneurosci.0590-05.2005
80. Bailey KR, Crawley JN (2009) Frontiers in neuroscience anxiety-related behaviors in mice. In: Buccafusco JJ (ed) Methods of behavior analysis in neuroscience. CRC Press/Taylor & Francis, Taylor & Francis Group, LLC, Boca Raton (FL)
81. Vermeer LM, Gregory E, Winter MK, McCarson KE, Berman NE (2015) Behavioral effects and mechanisms of migraine pathogenesis following estradiol exposure in a multibehavioral model of migraine in rat. Exp Neurol 263:8–16. https://doi.org/10.1016/j.expneurol.2014.09.011
82. Geng Y, Byun N, Delpire E (2010) Behavioral analysis of Ste20 kinase SPAK knockout mice. Behav Brain Res 208(2):377–382. https://doi.org/10.1016/j.bbr.2009.12.005
83. Takahashi A, Nishi A, Ishii A, Shiroishi T, Koide T (2008) Systematic analysis of emotionality in consomic mouse strains established from C57BL/6J and wild-derived MSM/Ms. Genes Brain Behav 7(8):849–858. https://doi.org/10.1111/j.1601-183X.2008.00419.x
84. Lister RG (1987) The use of a plus-maze to measure anxiety in the mouse. Psychopharmacology 92(2):180–185

85. Hogg S (1996) A review of the validity and variability of the elevated plus-maze as an animal model of anxiety. Pharmacol Biochem Behav 54(1):21–30
86. Lalonde R, Strazielle C (2008) Relations between open-field, elevated plus-maze, and emergence tests as displayed by C57/BL6J and BALB/c mice. J Neurosci Methods 171 (1):48–52. https://doi.org/10.1016/j.jneumeth.2008.02.003
87. Drummond GB, Paterson DJ, McGrath JC (2010) ARRIVE: new guidelines for reporting animal research. J Physiol 588(Pt 14):2517. https://doi.org/10.1113/jphysiol.2010.192260
88. Gates H, Mallon AM, Brown SD (2011) High-throughput mouse phenotyping. Methods 53 (4):394–404. https://doi.org/10.1016/j.ymeth.2010.12.017
89. Fuchs H, Gailus-Durner V, Adler T, Aguilar-Pimentel JA, Becker L, Calzada-Wack J, Da Silva-Buttkus P, Neff F, Gotz A, Hans W, Holter SM, Horsch M, Kastenmuller G, Kemter E, Lengger C, Maier H, Matloka M, Moller G, Naton B, Prehn C, Puk O, Racz I, Rathkolb B, Romisch-Margl W, Rozman J, Wang-Sattler R, Schrewe A, Stoger C, Tost M, Adamski J, Aigner B, Beckers J, Behrendt H, Busch DH, Esposito I, Graw J, Illig T, Ivandic B, Klingenspor M, Klopstock T, Kremmer E, Mempel M, Neschen S, Ollert M, Schulz H, Suhre K, Wolf E, Wurst W, Zimmer A, Hrabe de Angelis M (2011) Mouse phenotyping. Methods 53(2):120–135. https://doi.org/10.1016/j.ymeth.2010.08.006
90. Mandillo S, Tucci V, Holter SM, Meziane H, Banchaabouchi MA, Kallnik M, Lad HV, Nolan PM, Ouagazzal AM, Coghill EL, Gale K, Golini E, Jacquot S, Krezel W, Parker A, Riet F, Schneider I, Marazziti D, Auwerx J, Brown SD, Chambon P, Rosenthal N, Tocchini-Valentini G, Wurst W (2008) Reliability, robustness, and reproducibility in mouse behavioral phenotyping: a cross-laboratory study. Physiol Genomics 34(3):243–255. https://doi.org/10.1152/physiolgenomics.90207.2008
91. Rosen B, Schick J, Wurst W (2015) Beyond knockouts: the International Knockout Mouse Consortium delivers modular and evolving tools for investigating mammalian genes. Mamm Genome 26(9–10):456–466. https://doi.org/10.1007/s00335-015-9598-3
92. Karp NA, Meehan TF, Morgan H, Mason JC, Blake A, Kurbatova N, Smedley D, Jacobsen J, Mott RF, Iyer V, Matthews P, Melvin DG, Wells S, Flenniken AM, Masuya H, Wakana S, White JK, Lloyd KC, Reynolds CL, Paylor R, West DB, Svenson KL, Chesler EJ, de Angelis MH, Tocchini-Valentini GP, Sorg T, Herault Y, Parkinson H, Mallon AM, Brown SD (2015) Applying the ARRIVE Guidelines to an In Vivo Database. PLoS Biol 13(5):e1002151. https://doi.org/10.1371/journal.pbio.1002151
93. Kurbatova N, Mason JC, Morgan H, Meehan TF, Karp NA (2015) PhenStat: a tool kit for standardized analysis of high throughput phenotypic data. PLoS One 10(7):e0131274. https://doi.org/10.1371/journal.pone.0131274

Chapter 16

Cryobanking and Recovery of Genetically Modified Mice

Toru Takeo and Naomi Nakagata

Abstract

Cryobanking of sperm, oocytes, and embryos is a useful means to efficiently maintain mouse colonies without breeding live animals. Cryopreserved cells can be permanently stored in well-managed systems in liquid nitrogen tanks at −196 °C and quickly reanimated for use via in vitro fertilization and/or embryo transfer. Recent improvements of reproductive technology markedly enhanced the efficiency of recovering and producing animals using cryopreserved cells. The establishment of a cryobanking system will increase the performance of animal experiments, meet the principles of 3Rs (replacement, reduction, and refinement), and reduce labour and costs. In this chapter, we described the latest techniques of sperm cryopreservation, in vitro fertilization, and oocyte and two-cell embryo vitrification developed at the Center for Animal Resources and Development (CARD).

Key words Sperm cryopreservation, In vitro fertilization, Oocyte vitrification, Two-cell embryo vitrification, Embryo transfer

1 Introduction

Sperm cryopreservation is the first choice for the rapid preservation of mouse lines [1]. Epididymal sperm collected from the cauda epididymides of matured male mice are suspended in sugar-based cryoprotectant, which is loaded in freezing straws and preserved at −196 °C [2]. The cryopreserved sperm can be used for efficient fertilization using improved in vitro fertilization (IVF) systems featuring methyl-β-cyclodextrin (MBCD) and reduced glutathione (GSH) [3, 4].

To enhance the yield of oocytes used for IVF, the ultra-superovulation technique can be employed by the combined administration of inhibin antiserum (IAS) and equine chorionic gonadotropin (eCG) 48 h before human chorionic gonadotropin (hCG) administration [5, 6]. In addition, the oocytes can be cryopreserved before use [7]. After IVF, two-cell embryos can be preserved by embryo cryopreservation or used to produce animals via embryo transfer. In embryo cryopreservation, two-cell embryos are vitrified in cryoprotectants using 1 M dimethyl sulfoxide

Melissa A. Larson (ed.), *Transgenic Mouse: Methods and Protocols*, Methods in Molecular Biology, vol. 2066,
https://doi.org/10.1007/978-1-4939-9837-1_16, © Springer Science+Business Media, LLC, part of Springer Nature 2020

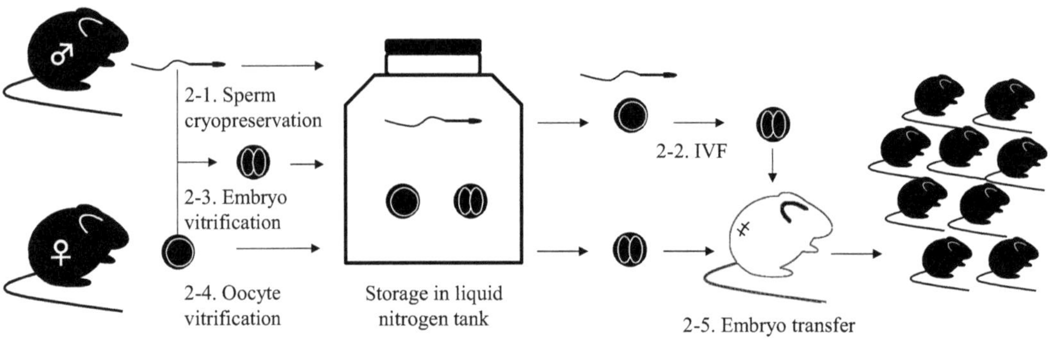

Fig. 1 Cryobanking system of sperm, oocytes, and embryos harvested from genetically modified mice

(DMSO) and DAP213 (2 M DMSO, 1 M acetamide, and 3 M propylene glycol) and stored at −196 °C [8]. In embryo transfer, two-cell embryos are transferred into the oviducts of pseudopregnant female mice and offspring are delivered 19.5 days after embryo transfer [9].

Thus, we can efficiently preserve mouse cell lines via cryopreservation techniques and reanimate them using related reproductive techniques (Fig. 1). The establishment of a cryobanking system in your institute will help refine the performance of animal experiments using GM mice and achieve the cost-effective management of animal facilities.

2 Materials

2.1 Equipment, Mice, and Plasticware

1. Male mice to be cryopreserved.
2. Female oocyte donor mice of chosen strain (e.g., C57BL/6).
3. Female recipient mice of chosen strain (e.g., CD-1).
4. Vasectomized male mice, strain unimportant (e.g., CD-1).
5. Surgical instruments: No. 5 forceps, 4″ straight scissors, microspring scissors, bulldog clamp, semen straw cutter, dissecting needles, wound clip applicator (Fig. 2).
6. Stereomicroscope with magnification 8–80×.
7. Hot plate.
8. 0.25 mL Clear straws (10 per male).
9. Straw connector: 1.0 mL syringe, three-way stopcock, plastic tube, silicone tube.
10. Impulse heat sealer.
11. Cryogenic labels for straws and tubes.
12. TLS PC Link thermal transfer printer or similar thermal printer.

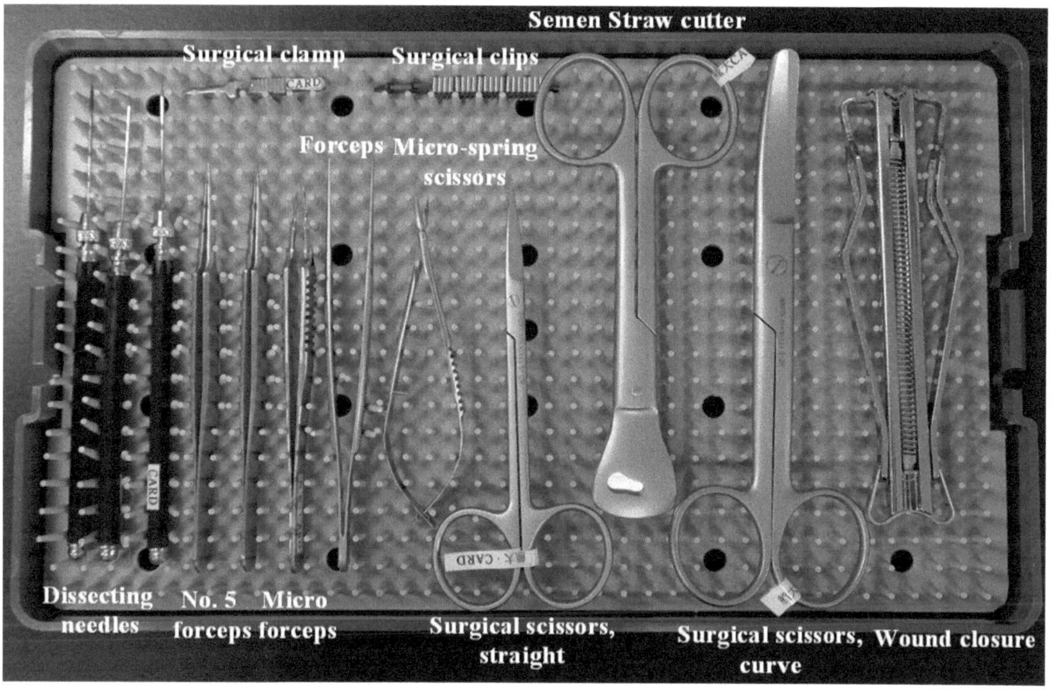

Fig. 2 Surgical instruments for mouse reproductive techniques

13. 35 mm × 10 mm Plastic Petri dishes.
14. Micropipettors and tips.
15. Freezing canister: acrylic bar, a 50-ml disposable syringe, Styrofoam float.
16. Liquid nitrogen and tank.
17. Cryogenic storage cassettes for straws.
18. CO_2 incubator.
19. Water bath.
20. Cryotubes.
21. Block cooler, such as Nalgene Labtop Cooler.

2.2 Media

1. Sperm cryoprotectant (CPA): Dissolve 3.6 g of raffinose pentahydrate, 0.6 g of skim milk, and 0.292 g of L-glutamine in 20 mL of distilled water at 60 °C in a 50-mL centrifuge tube. Incubate the tube in a water bath at 60 °C for 90 min, vortexing every 30 min. Transfer 1.0 mL aliquots of the solution into 1.5 mL microcentrifuge tubes. Centrifuge the solution at 10,000 × *g* for 60 min. Carefully collect the supernatant using a micropipette (*see* **Note 1**). Filtrate the supernatant using a 0.22-μm disposable filter and use it as sperm cryoprotectant (modified R18S3: mR18S3). Transfer the solution of

mR18S3 into 1.0 mL ampules, fill the space above the liquid with nitrogen gas, and store at room temperature (use within 3 months).

2. Sperm preincubation medium: 697.6 mg NaCl, 35.6 mg KCl, 25.1 mg $CaCl_2 2H_2O$, 100 mg glucose, 29.3 mg $MgSO_4 7H_2O$, 16.2 mg KH_2PO_4, 5.5 mg sodium pyruvate, 210.6 mg $NaHCO_3$, 7.5 mg penicillin G, 5 mg streptomycin, 100 mg polyvinyl alcohol, 98.3 mg methyl-β-cyclodextrin. Add 80 mL sterile water and stir. Bring final volume to 100 mL and filter. Aliquot into ampules and store at 4°C. Alternatively, FERTIUP® Mouse Sperm Preincubation Medium can be purchased commercially.
3. Modified human tubal fluid: 593.8 mg NaCl, 35 mg KCl, 57 mg $CaCl_2$, 50 mg glucose, 4.9 mg $MgSO_4 7H_2O_2$, 5.4 mg KH_2PO_4, 3.5 mg sodium pyruvate, 0.34 mL 60% sodium lactate, 210 mg $NaHCO_3$, 7.5 mg penicillin G, 5 mg streptomycin, 400 mg bovine serum albumin, 0.04 mL 0.5% phenol red. Add 80 mL sterile water and stir. Bring final volume to 100 mL and filter. Alternatively, CARD MEDIUM® can be purchased commercially.
4. KSOM/AA culture medium: 555 mg NaCl, 18.5 mg KCl, 4.75 mg KH_2PO_4, 4.95 mg $MgSO_4 7H_2O$, 25 mg $CaCl_2 2H_2O$, 210 mg $NaHCO_3$, 3.6 mg glucose, 2.2 mg sodium pyruvate, 0.174 mL DL-lactic acid, sodium salt, 100 μL 10 mM EDTA, 5 mg streptomycin, 6.3 mg penicillin, 0.1 mL 0.5% phenol red, 14.6 mg L-glutamine, 1 mL MEM essential amino acids, 0.5 mL MEM nonessential amino acids, 100 mg bovine serum albumin. Add 80 mL sterile water and stir. Bring final volume to 100 mL and filter. Alternatively, KSOM may be purchased commercially.
5. Inhibin antiserum (IAS) and equine chorionic gonadotropin (eCG) reconstituted to 75 IU/mL in sterile PBS. Aliquot and store at −20°C. Alternatively, CARD HyperOva can be purchased commercially.
6. Human chorionic gonadotropin (hCG) reconstituted to 75 IU/mL in sterile PBS. Aliquot and store at −20°C.
7. PB1 medium: 800 mg NaCl, 20 mg KCl, 12 mg $CaCl_2$, 20 mg KH_2PO_4, 10 mg $MgCl_2{\cdot}6H_2O$, 115 mg Na_2HPO_4, 36 mg sodium pyruvate, 100 mg glucose, 7.5 mg penicillin, 5 mg streptomycin, 300 mg bovine serum albumin. Add 80 mL sterile water and stir. Bring final volume to 100 mL and filter.
8. 1 M Dimethyl sulfoxide (DMSO): 7.813 mL DMSO, 92.2 mL PB1. Filter. Alternatively, 1 M DMSO can be purchased commercially.

9. DAP213: Prepare 2.3088 mL PB1, 3.1252 mL DMSO, 4.556 mL propylene glycol and mix. Prepare 1181.4 mg acetamide in 10 mL PB1 and dissolve. Combine equal volumes of both solutions to create DAP213. Alternatively, DAP213 can be purchased commercially.
10. 0.25 M Sucrose solution: 1711.5 mg sucrose in 20 mL of PB1.
11. 1% Hyaluronidase.
12. Fetal bovine serum.
13. Paraffin liquid.

3 Methods

3.1 Sperm Cryopreservation

For sperm cryopreservation, three matured male mice (3–6 months old), sperm cryoprotectant, and various materials are needed. The protocol is based on our previous work [1, 2].

1. Place a 60-μL drop of CPA in a 35-mm culture dish and cover with paraffin liquid.
2. Place another 60 μL aliquot of CPA on top of the first drop (total, 120 μL CPA/two epididymides) to make a tall drop (*see* **Note 2**).
3. Euthanize a matured male mouse (3–6 months old) by cervical dislocation and aseptically collect the two cauda epididymides (Fig. 3a).
4. Place the cauda epididymides on sterile filter paper. Completely remove adipose tissue and blood from the cauda epididymides under a microscope using a pair of No. 5 forceps and surgical scissors (straight).
5. Transfer the cauda epididymides into the CPA drop. Make five or six deep cuts in the epididymides using microspring scissors (5 mm blades) under the microscope (Fig. 3b).
6. Place the dish on a hot plate at 37 °C for 3 min. Gently swirl the dish every minute to uniformly disperse the sperm in the CPA (Fig. 3c).
7. During equilibration of the sperm in the CPA, attach the straw connector to the straw (*see* **Note 3**). Carefully aspirate 0.1 mL of modified human tubal fluid (mHTF) and 15 mm of air into the straw. Ten straws can be prepared from the two epididymides of a single male.
8. Using a micropipette, place ten 10 μL aliquots of sperm in a plastic culture dish.
9. Aspirate one 10 μL drop of sperm and 15 mm of air into each of the 10 straws (Fig. 3d).

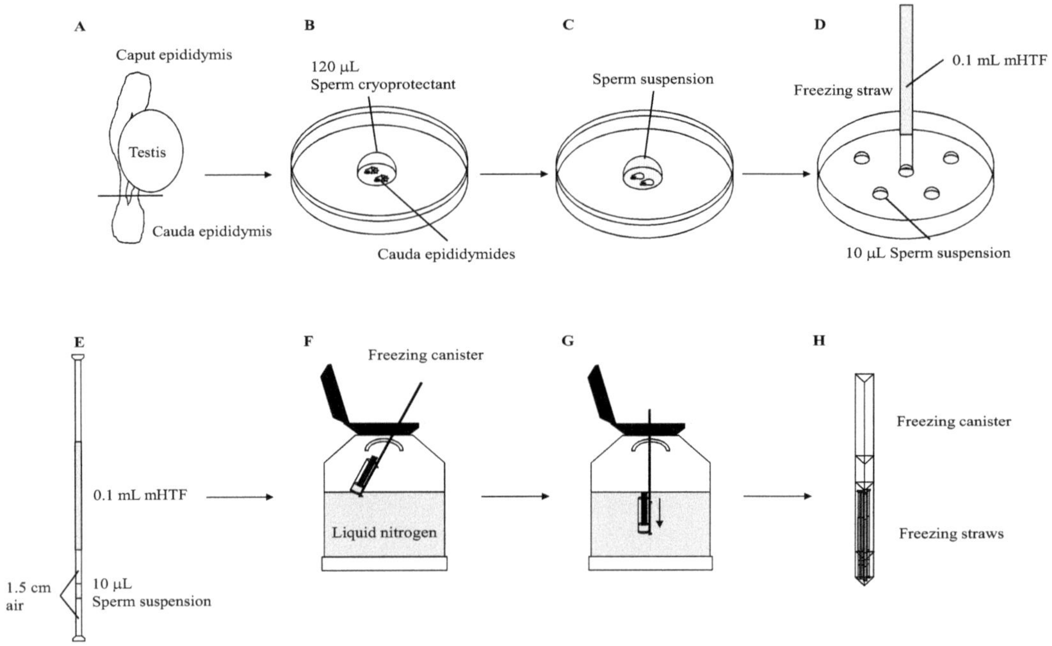

Fig. 3 Sperm cryopreservation

10. Carefully seal both ends of the straw using the impulse sealer (Fig. 3e) (*see* **Note 4**).
11. Place the sealed straws into the freezing canister and float the canister on the surface of liquid nitrogen in a liquid nitrogen tank for 10 min (Fig. 3f).
12. After 10 min, completely immerse the freezing canister in the liquid nitrogen (Fig. 3g).
13. Transfer the straws to a triangular cassette that was precooled in liquid nitrogen and then immerse the cassette in the tank for long-term storage (Fig. 3h) (*see* **Note 5**).

3.2 Sperm Thawing

1. Place a 90-μL drop of sperm preincubation medium in a 35-mm culture dish and cover it with paraffin liquid (sperm preincubation drop). Equilibrate the medium in an incubator (37 °C, 5% CO_2) for 30 min before use.
2. Remove a frozen straw of sperm from the liquid nitrogen tank (Fig. 4a) and hold it in the air for 5 s.
3. Place the frozen straw in the floating container composed of Styrofoam and a 50-mL plastic centrifuge tube and incubate it in a 37 °C water bath for 10 min to thaw (Fig. 4b).
4. Remove the straw from the water bath and dry with a paper towel.

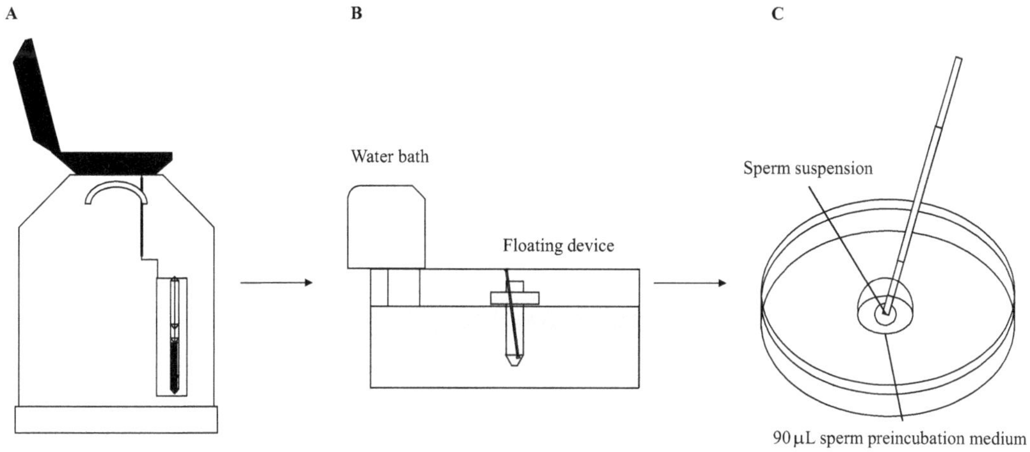

Fig. 4 Thawing of cryopreserved sperm

5. Cut the straw between the mHTF medium and the seal using a semen straw cutter or straight scissors. Then, insert the straw into the straw connector and turn the stopcock to release internal pressure. Finally, cut the sealed end of the straw closest to the sperm.
6. Push the plunger of the straw connector and transfer only the sperm suspension (10 μL) to the 90 μL drop of sperm preincubation medium in the culture dish (Fig. 4c).
7. Place the dish in a humidified incubator (37 °C, 5% CO_2) for 30 min before insemination.

3.3 In Vitro Fertilization

1. Inject 0.1 mL of IAS and 3.75 IU of eCG (IASe), 0.2 mL of CARD HyperOva, or 7.5 IU of eCG into female mice (4-week-old or >12-week-old) at 4–6 PM (*see* **Note 6**).
2. Inject 7.5 IU of hCG into the female mice 48 h after the IASe injection at 4–6 PM.
3. Collect oocytes 15–17 h after the hCG injection.
4. Place a 90 μL drop of fertilization medium (mHTF with 1.0 mM GSH or CARD MEDIUM®) in a plastic culture dish and cover it with paraffin liquid (fertilization drop). Equilibrate the medium in a humidified incubator at 37 °C in 5% CO_2 for 10 min (*see* **Note 7**).
5. Place four 80 μL drops of mHTF in a plastic culture dish and cover the dish with paraffin liquid (washing drops). Incubate this washing dish in a humidified incubator (37 °C, 5% CO_2) for at least 30 min.
6. Fifteen to seventeen hours after the hCG injection, euthanize the superovulated mice by cervical dislocation and quickly collect oviducts.

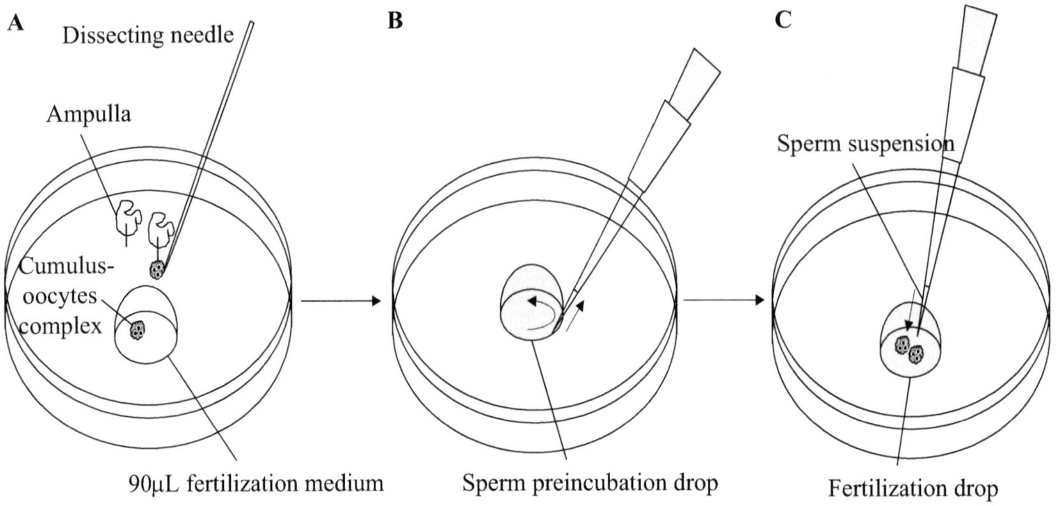

Fig. 5 In vitro fertilization

7. Place the oviducts on sterile filter paper. Remove any blood or fluid from the tissue under a microscope.
8. Immerse the oviducts in the paraffin liquid contained in the culture dish with the fertilization drop.
9. Under a microscope, hold each oviduct with No. 5 forceps and gently tear the swollen ampulla of the oviduct with a dissecting needle (Fig. 5a). After tearing the swollen ampulla, wait for the cumulus–oocyte complexes (COCs) to be released into the paraffin liquid.
10. Introduce the COCs into the drop of fertilization medium using the dissecting needle (2–4 COCs/drop). Place the dish in a humidified incubator (37 °C, 5% CO_2) and incubate for 30–60 min before insemination (*see* **Note 8**).
11. After sperm preincubation in the incubator, aspirate a 10-μL aliquot containing motile sperm from the edge of the sperm preincubation drop using a pipette tip for insemination (Fig. 5b) (*see* **Note 9**).
12. Add the sperm suspension to the fertilization drop containing COCs (Fig. 5c). Place the culture dish in a humidified incubator (37 °C, 5% CO_2).
13. After incubating for 3 h, wash the oocytes three times in the 80 μL drops of fresh mHTF in the washing dish (*see* **Note 10**).
14. Six hours after insemination, observe oocytes in the third drop of mHTF and remove any parthenogenetic oocytes.
15. After overnight culture of the oocytes, transfer only the two-cell embryos to the fourth drop of mHTF in the washing dish. These two-cell embryos are ready for embryo cryopreservation or embryo transfer to pseudopregnant females.

3.4 Two-Cell Embryo Vitrification

Two-cell embryos can be preserved by vitrification. In this section, we describe a vitrification method using 1 M DMSO and DAP213 (2 M DMSO, 1 M acetamide, and 3 M propylene glycol) [8].

1. Filtrate the 1 M DMSO in PB1 solution and put four drops of the solution (100 μL/drop) into a dish (*see* **Note 11**).
2. Place a group of embryos into one of the four drops to rinse off the collection medium.
3. Divide the rinsed embryos equally between the other drops. These aliquots will eventually be transferred to a cryotube (Fig. 6a) (*see* **Note 12**).
4. Using a 20-μL pipette and a gel-loading tip, transfer the embryos contained within 5 μL of 1 M DMSO solution into a cryotube (Fig. 6b) (*see* **Note 13**).
5. Once transferred, put the cryotube into the block cooler at 0 °C and wait for 5 min (*see* **Note 14**).
6. Add 45 μL of cryoprotective solution composed of 2 M DMSO, 1 M acetamide, and 3 M propylene glycol in PB1 (DAP213) at 0 °C into the cryotube and equilibrate for 5 min in the block cooler at 0 °C (Fig. 6c) (*see* **Note 15**).
7. Quickly set the cryotubes on a storage cane and plunge the samples directly into liquid nitrogen (Fig. 6d).
8. Store the samples under a liquid phase in the liquid nitrogen tank (Fig. 6e).

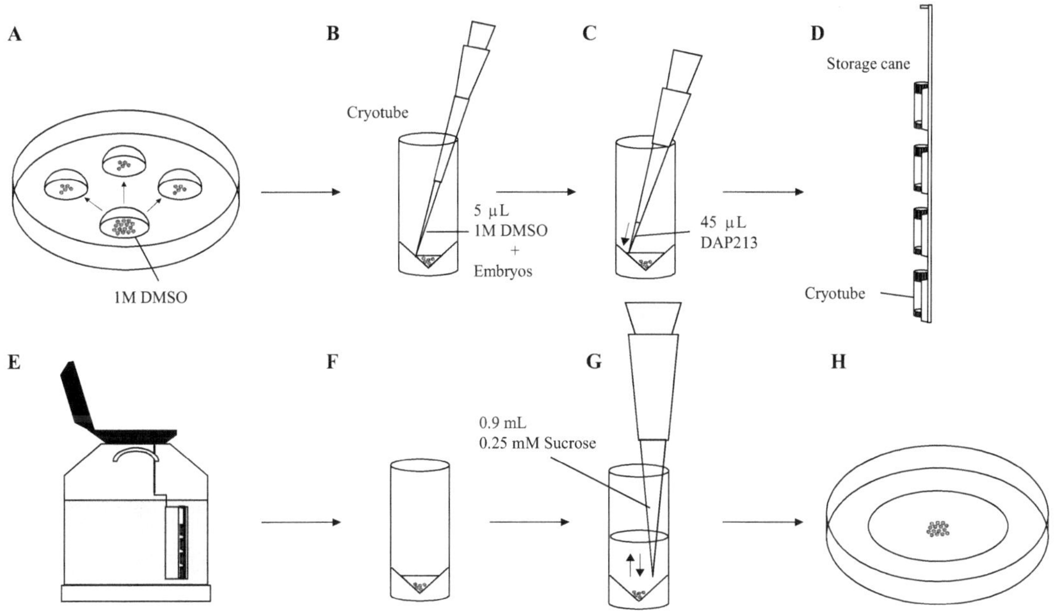

Fig. 6 Embryo vitrification and warming

3.5 Embryo Warming

1. Remove the required sample from the liquid nitrogen and open the cap of cryotube.
2. Discard any liquid nitrogen in the tube and allow the tube to stand at room temperature for 30 s (Fig. 6f).
3. Add 0.9 mL of 0.25 M sucrose solution preheated at 37 °C to the cryotube and warm the sample quickly via pipetting (Fig. 6g) (*see* **Note 16**).
4. Once warmed, transfer the contents of the cryotube into a plastic dish (Fig. 6h).
5. Place 0.4–0.5 mL of 0.25 M sucrose into the cryotube and transfer the contents into the plastic dish (*see* **Note 17**).
6. Aspirate the embryos from the liquid, carefully transfer them into a drop of KSOM/AA (washing dish), and then store them in an incubator (37 °C, 5% CO_2 in air).
7. After 10 min, wash the embryos twice with fresh KSOM/AA (washing dish).

3.6 Oocyte Vitrification

This vitrification method is applicable to the cryopreservation of oocytes. After removing cumulus cells, the oocytes can be preserved and used for IVF after warming [7].

1. Collect COCs from superovulated female mice and introduce them into a 200-μL drop of mHTF (Fig. 7a) (*see* **Note 18**).
2. Add 20 μL of 1% hyaluronidase to the drop of mHTF containing the COCs and store the dish in an incubator (37 °C, 5% CO_2 in air) for 1 min (Fig. 7b).
3. Promptly collect and transfer the oocytes into an 80-μL drop of mHTF and then wash them in the drops in the washing dish (Fig. 7c) (*see* **Note 19**).
4. Transfer the oocytes into the first drop in the fetal bovine serum (FBS) dish to rinse. Then, transfer them into the second drop to incubate (37 °C, 5% CO_2 in air) for 10 min (Fig. 7d) (*see* **Note 20**).

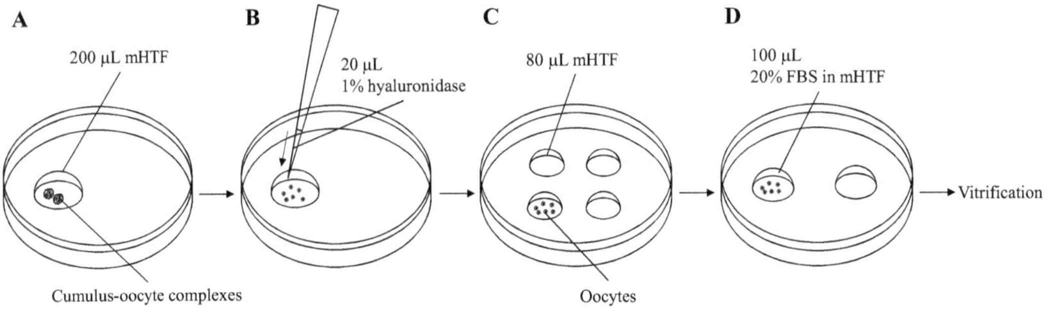

Fig. 7 Removal of cumulus cells from oocytes for oocyte vitrification

5. The oocytes can be vitrified using the simple vitrification method for two-cell embryos after removing cumulus cells and culturing them in a drop containing FBS.

3.7 Oocyte Warming

1. The warming method is the same as that for two-cell embryos, except mHTF drops are used for washing warmed oocytes.
2. The vitrified–warmed oocytes can be used for in vitro fertilization using fresh, frozen–thawed, and cold temperature-transported spermatozoa (*see* **Note 21**).

3.8 Embryo Transfer

Two-cell embryos develop to pups at 19.5 days after embryo transfer. In this section, we describe embryo transfer through the wall of the ampulla of pseudopregnant female mice [9].

1. Select matured female mice (CD-1 strain) at the proestrus stage by observing the appearance of the vagina (gaping with reddish pink and wet tissues).
2. Mate the female mice with vasectomized male mice at 3–5 PM.
3. In the next morning, observe the females for the formation of vaginal plugs and use those with plugs as recipients of embryo transfer.
4. Anesthetize the recipient mouse.
5. Remove the ovary, oviduct, and part of the uterine horn per the conventional procedure.
6. Clip a surgical clamp onto the fat pad, which is attached to the ovarian bursa.
7. Observe the oviduct under a stereomicroscope and confirm the position of the infundibulum and ampulla using the tip of a set of forceps or by changing the position of the surgical clamp.
8. Position the oviduct by changing the position of the surgical clamp and the mouse (*see* **Note 22**).
9. Add a 200-μL drop of KSOM/AA in a dish (without liquid paraffin) and introduce 20 embryos into the drop (Fig. 8a).
10. Aspirate air and medium in alternate intervals of 2–3 mm into a glass capillary in preparation for embryo transfer. Draw 10 embryos into the glass capillary (Fig. 8b).
11. Using a pair of microforceps and microspring scissors, dissect the wall of the oviduct (∗) between the infundibulum and ampulla (Fig. 8c).
12. Insert the tip of the capillary containing the embryos into the slit and then push the capillary further into the slit toward the ampulla (Fig. 8d).
13. Use the forceps to hold the portion of the oviduct into which the capillary was inserted.

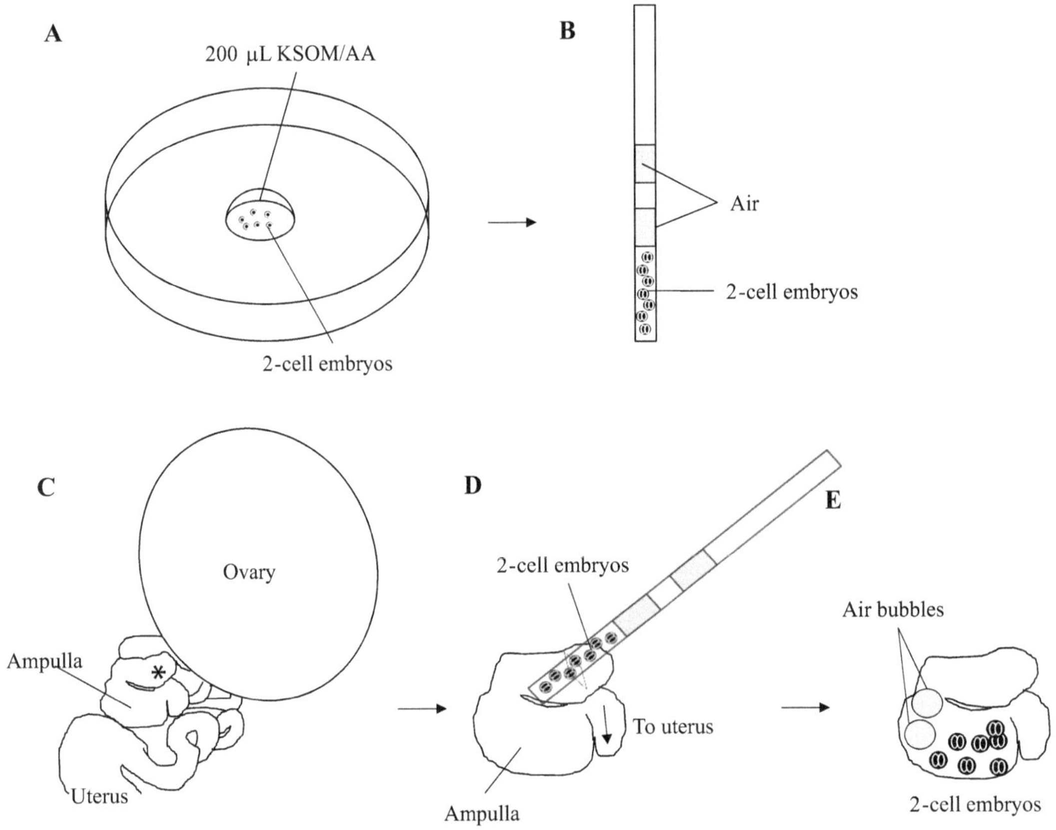

Fig. 8 Embryo transfer through the wall of the oviduct

14. Expel the embryos and 2–3 of the air bubbles into the ampulla (Fig. 8e) (*see* **Note 23**).
15. Withdraw the capillary gently from the slit.
16. Push the ovary, oviduct, and uterine horn back into the abdomen and close the wound using surgical clips.
17. Repeat the process to transfer the remaining 10 embryos into the other oviduct as described previously.
18. Maintain the mice warm on a 37 °C warming plate until they recover from the effects of the anesthesia.

4 Notes

1. If the supernatant is not clear, then centrifuge the solution until it becomes clear.
2. This results in a more uniform sperm suspension by mixing and reduces the amount of residual sperm remaining in the dish after collecting the sperm.

3. Information about the preserved mouse line (strain name, project ID, and preservation date) is indicated on the straws by the printer labels.
4. Confirm that both ends of the straw are well sealed to prevent liquid nitrogen from entering during storage, which could result in the straw exploding during warming.
5. To check the quality of sperm, transfer a 1.0-μL aliquot of the sperm suspension from the CPA drop into a 100-μL drop of mHTF covered with paraffin liquid in a plastic culture dish. Observe sperm motility under the microscope after incubating the sperm in a CO_2 incubator for 10 min.
6. Increasing the efficiency of superovulation is important for reducing the number of oocyte donors. Recently, we developed a novel superovulation technique using the combined administration of IAS and eCG (IASe method), which more effectively increases the number of ovulated oocytes than eCG alone [5, 6].
7. Mouse sperm, especially those derived from the C57BL/6 strain, exhibit decreased motility and fertilizing ability after freezing and thawing [10]. The fertilization rate of frozen–thawed sperm can be recovered via preincubation in TYH containing MBCD and fertilization in mHTF containing GSH [3, 4].
8. Perform all operations—from sacrificing the female mice to introducing the COCs into the fertilization medium—in the shortest time possible (preferably within 30 s). If performing this process alone, sacrifice one mouse at a time and swiftly remove its oviducts before proceeding to the next mouse.
9. Carefully collect motile sperm from the periphery of the drop. It is possible to collect four 10 μL aliquots of sperm suspension per drop.
10. If many spermatozoa are attached to the zona pellucida of the oocytes, they can be removed by pipetting up and down (using a 20-μL micropipette and a tip) 20–30 times in the fertilization drop before washing.
11. One drop will be used to wash the embryos taken from the collection medium, whereas the others will be used to hold the washed embryos.
12. For example, if one were to collect 120 embryos and vitrify them in 40-embryo aliquots, the embryos would first be placed together in the solution drop for rinsing and then divided equally among the remaining three drops.

13. If the embryos are pushed together in the center of the drop, it is easy to capture all of them in 5 μL of the 1 M DMSO solution.
14. It is possible to keep the cryotubes in the block cooler at 0 °C for longer than 5 min (<20 min).
15. Do not fasten the caps of cryotube too tightly after adding the DAP213, or they will be too difficult to quickly discard liquid nitrogen in the tube when samples are warmed and recovered.
16. When pipetting, take care not to generate large amounts of bubbles and physically damage embryos by pipetting too quickly.
17. This further dilutes the cryoprotectant and ensures that all of the embryos have been transferred.
18. Be sure to perform all operations, from sacrificing the female and removing her oviducts to introducing the COCs into a drop of mHTF (oocyte dish), in the shortest time possible (within 30 s). Moreover, when performing this process alone, do not sacrifice multiple mice at once; instead, sacrifice one mouse and swiftly remove its oviducts before moving on to the next mouse.
19. If some cumulus cells become attached to the zona pellucida of the oocytes, they can be removed by manipulating the glass capillary.
20. FBS can prevent zona hardening in the oocyte during vitrifying and warming.
21. There are three different methods for preparing CARD MEDIUM depending on whether in vitro fertilization will be performed using fresh, frozen–thawed, or cold temperature-transported spermatozoa [7].
22. The folds in the oviduct vary between each mouse; thus, look closely and adjust the position of the oviduct to increase the ease of manipulation. If you are left-handed, position the oviduct to allow yourself to easily perform the procedure with your left hand.
23. If performed successfully, you should be able to see air bubbles through the wall of the ampulla. If you cannot expel the embryos and air bubbles into the oviduct, pull the capillary out slightly from the slit and expel them again.

Acknowledgments

We thank our staff, namely Kiyoko Yamashita, Tomoko Kondo, Yukie Haruguchi, Yumi Takeshita, Yuko Nakamuta, Tomoko Umeno, and Eri Ishida, and our students Hidetaka Yoshimoto and Ayumi Mukunoki for technical support. This work was partially supported by grants of the National Bioresource Project from the Ministry of Education, Culture, Sports, Science and Technology of Japan and Research on Development of New Drugs from the Japan Agency for Medical Research and Development.

References

1. Nakagata N (2000) Cryopreservation of mouse spermatozoa. Mamm Genome 11 (7):572–576
2. Takeo T, Nakagata N (2010) Combination medium of cryoprotective agents containing L-glutamine and methyl-{beta}-cyclodextrin in a preincubation medium yields a high fertilization rate for cryopreserved C57BL/6J mouse sperm. Lab Anim 44(2):132–137. https://doi.org/10.1258/la.2009.009074
3. Takeo T, Hoshii T, Kondo Y, Toyodome H, Arima H, Yamamura K, Irie T, Nakagata N (2008) Methyl-beta-cyclodextrin improves fertilizing ability of C57BL/6 mouse sperm after freezing and thawing by facilitating cholesterol efflux from the cells. Biol Reprod 78 (3):546–551. https://doi.org/10.1095/biolreprod.107.065359
4. Takeo T, Nakagata N (2011) Reduced glutathione enhances fertility of frozen/thawed C57BL/6 mouse sperm after exposure to methyl-beta-cyclodextrin. Biol Reprod 85 (5):1066–1072. https://doi.org/10.1095/biolreprod.111.092536
5. Takeo T, Nakagata N (2015) Superovulation using the combined administration of inhibin antiserum and equine chorionic gonadotropin increases the number of ovulated oocytes in C57BL/6 female mice. PLoS One 10(5): e0128330. https://doi.org/10.1371/journal.pone.0128330
6. Takeo T, Nakagata N (2016) Immunotherapy using inhibin antiserum enhanced the efficacy of equine chorionic gonadotropin on superovulation in major inbred and outbred mice strains. Theriogenology 86(5):1341–1346. https://doi.org/10.1016/j.theriogenology.2016.04.076
7. Nakagata N, Takeo T, Fukumoto K, Kondo T, Haruguchi Y, Takeshita Y, Nakamuta Y, Matsunaga H, Tsuchiyama S, Ishizuka Y, Araki K (2013) Applications of cryopreserved unfertilized mouse oocytes for in vitro fertilization. Cryobiology 67(2):188–192. https://doi.org/10.1016/j.cryobiol.2013.06.011
8. Nakao K, Nakagata N, Katsuki M (1997) Simple and efficient vitrification procedure for cryopreservation of mouse embryos. Exp Anim 46(3):231–234
9. Nakagata N (1992) Embryo transfer through the wall of the fallopian tube in mice. Jikken Dobutsu 41(3):387–388
10. Nishizono H, Shioda M, Takeo T, Irie T, Nakagata N (2004) Decrease of fertilizing ability of mouse spermatozoa after freezing and thawing is related to cellular injury. Biol Reprod 71 (3):973–978. https://doi.org/10.1095/biolreprod.103.024422

Chapter 17

Simple Transportation of Genetically Engineered Mice via Cold Storage Techniques

Hidetaka Yoshimoto, Toru Takeo, and Naomi Nakagata

Abstract

Live genetically modified (GM) mice are often shipped in transportation cages. However, the shipment of live mice is sometimes associated with difficulties. Recently, we developed an alternative means to transport GM mice via cold storage techniques for sperm and embryos. Cold storage of the cauda epididymis, the male reproductive organ for storing mature sperm, can maintain the fertility of sperm in cold preservation medium for 10 days. The sperm can then be used to produce embryos or pups via in vitro fertilization and embryo transfer. Conversely, cold-stored two-cell embryos maintain developmental ability for 4 days. The embryos can be used to produce pups via embryo transfer. Cold transport techniques are being increasingly employed to create local and global networks between research institutes for the shipment of GM mice. In this text, we described a cold transport technique for sperm and two-cell embryos in mice.

Key words Transportation, Cold storage, Cauda epididymis, Two-cell embryo

1 Introduction

The shipment of fresh sperm and embryos using cold storage techniques can be employed to transport a genetically modified (GM) mouse line. These techniques avoid potential problems such as the expansion of microbiological contamination, animal escape or death, the legal restrictions of GM organisms, and high shipment costs associated with the transport of live animals.

The cold storage of cauda epididymides, which are male reproductive organs that store mature sperm, is a simple technology [1]. Transported sperm can be used to perform in vitro fertilization (IVF) and obtain offspring via embryo transfer. Recently, we demonstrated that Lifor preservation medium containing dimethyl sulfoxide and quercetin can maintain the fertilizing ability of cold-stored sperm for 10 days [1–3]. The cold-stored sperm exhibited high fertilization rates in IVF systems using methyl-beta-cyclodextrin (MBCD) and reduced glutathione (GSH) [1, 2, 4, 5]. In addition, the transported sperm are readily available for IVF with

Melissa A. Larson (ed.), *Transgenic Mouse: Methods and Protocols*, Methods in Molecular Biology, vol. 2066,
https://doi.org/10.1007/978-1-4939-9837-1_17, © Springer Science+Business Media, LLC, part of Springer Nature 2020

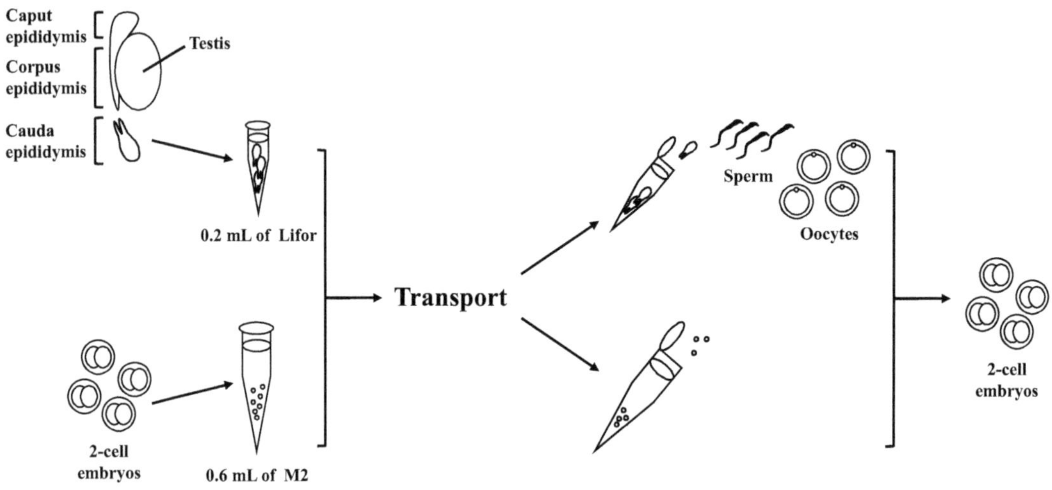

Fig. 1 Shipment of GM mice using the cold-storage technique of sperm and two-cell embryos

vitrified–warmed oocytes [6]. Furthermore, supplementation of Lifor with sphingosine 1-phosphate boosts its protective ability and makes it possible to cryopreserve sperm after cold storage for 3 days [7].

Cold storage of two-cell embryos is another option for fresh embryos derived from IVF or cryopreserved embryos derived from GM mice [8, 9]. Previously, we demonstrated that M2 medium maintained the developmental ability of fresh or vitrified–warmed two-cell embryos held at 4 °C for 3 days [9]. Recently, we discovered that *N*-acetyl cysteine (NAC) prolonged the cold storage period by 4 days by restoring the level of intracellular GSH [10].

Using cold transport techniques for sperm and two-cell embryos, we can efficiently receive and ship GM mice via our mouse bank system (Fig. 1). We strongly believe that the cold transport method will be standardized for the transport of GM mice between institutes and/or commercial companies.

2 Materials

2.1 Cold Storage of Cauda Epididymides

1. Matured male mice (>12-week old).
2. Microscissors.
3. Pair of No. 5 forceps.
4. Surgical clip (Autoclip 9 mm).
5. Surgical clip applicator (Mik-Ron Autoclip Applier).
6. Lifor preservation solution (Lifeblood Medical Co, Ltd) containing dimethyl sulfoxide and quercetin.

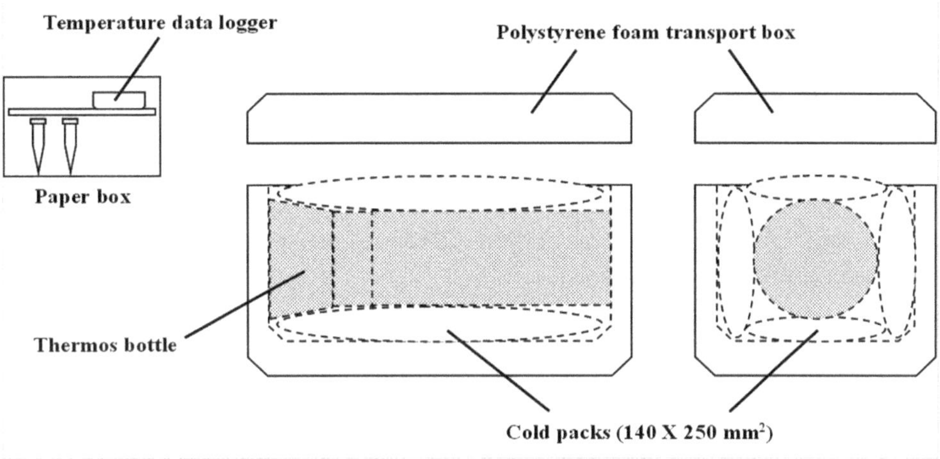

Fig. 2 CARD Cold Transport kit

7. CARD cold transport kit (Fig. 2).
   (a) Thermos bottle.
   (b) Paper box (in which a 0.2-mL tube can stand).
   (c) Cotton wool.
   (d) Cold packs (60 × 180 mm^2 and 140 × 250 mm^2).
   (e) Polystyrene foam transport box.
   (f) Temperature data logger (e.g., Thermochron iButton).

2.2 Preparation of Sperm for IVF

1. Sperm preincubation medium (TYH + 0.75 mM MBCD or FERTIUP® preincubation medium, Cosmo Bio).
2. Fertilization medium (mHTF + 0.5 mM GSH or CARD MEDIUM, Cosmo Bio).
3. mHTF (for washing the epididymides).
4. Plastic dishes.
5. Microspring scissors.
6. No. 5 forceps.
7. Filter paper.
8. Humidified incubator (37 °C, 5% CO_2, 95% air).

2.3 Cold Storage of Two-Cell Embryos

1. Two-cell embryos (adaptable for fresh and frozen/thawed embryos).
2. Plastic dishes.
3. Cold storage medium (M2 medium + 1.5 mM NAC).
4. 0.5-mL plastic tube.
5. CARD Cold Transport Kit (Cosmo Bio).

2.4 Collection of Cold-Stored Two-Cell Embryos

1. Gel-loading tip.
2. KSOM/AA (for washing the embryos).

3 Methods

3.1 Cold Storage of Cauda Epididymides [1]

1. All items in the CARD cold transport kit should be maintained at 4–8 °C.
2. Euthanize three male mice via cervical dislocation and collect cauda epididymides.
3. Transfer cauda epididymides to 0.2 mL of Lifor containing dimethyl sulfoxide and quercetin in 0.2-mL plastic tubes (3 epididymides/tube).
4. Place the tubes in a paper box with a temperature data logger and a sheet of cotton wool.
5. Place the box in a Thermos bottle with two cold packs (60×180 mm^2).
6. Place the bottle in a polystyrene foam transport box with four cold packs (140×250 mm^2).
7. Seal the lid of the polystyrene foam transport box using packing tape.
8. Maintain the polystyrene foam transport box at 4–8 °C until transport.

3.2 IVF Using Cold-Stored Sperm

1. After cold storage, collect the cauda epididymides from the storage medium and remove the medium gently with filter paper.
2. Wash the epididymides three times in three drops of mHTF.
3. Cut the duct of each cauda epididymis using microspring scissors and transfer suspensions of sperm from the cauda epididymides into sperm preincubation medium.
4. Incubate the sperm for 60 min at 37 °C with 5% CO_2.
5. Euthanize female mice with superovulation treatment via cervical dislocation and collect oviducts.
6. Place the oviducts in liquid paraffin on fertilization dish.
7. Collect and introduce cumulus–oocytes complexes (COCs) into the fertilization drop (200 μL fertilization drop) using a dissecting needle.

8. Add a 3-μL aliquot of preincubated sperm to the fertilization drop containing COCs. Note: Adjust the volume of the sperm suspension to optimize the final concentration of motile sperm to at least 400 sperm/μL in the fertilization drop.
9. Three hours after insemination, wash the oocytes in drops of mHTF.

3.3 Cold Storage of Two-Cell Embryos [10]

1. Cold packs ($60 \times 180\ mm^2$) and the Thermos bottle must be maintained at room temperature. Note: Cold packs ($140 \times 250\ mm^2$) and the polystyrene foam transport box must be maintained at 4–8 °C.
2. After IVF, transfer the two-cell embryos to a 100-μL drop of M2 medium.
3. Place the embryos into a 0.5 mL tube containing 600 μL of M2 medium (40 embryos/tube).
4. Place the tubes in a paper box with a temperature data logger and a sheet of cotton wool.
5. Place the box in a Thermos bottle with two cold packs ($60 \times 180\ mm^2$).
6. Place the bottle in a polystyrene foam transport box with four cold packs ($140 \times 250\ mm^2$).
7. Seal the lid of the polystyrene foam transport box using packing tape.
8. Keep the polystyrene foam transport box at 4–8 °C until transport.

3.4 Collection of Cold-Stored Two-Cell Embryos

1. Put three drops (100 μL/drop) of KSOM/AA into a dish and cover then with liquid paraffin (washing dish).
2. Incubate the dish for at least 30 min at 37 °C with 5% CO_2.
3. After cold storage, collect the upper layer of 200 μL of cold storage medium from the tube using a gel-loading tip and then transfer the aliquot to the edge of the plastic dish.
4. Carefully retrieve the entire aliquot of cold storage medium containing the embryos from the bottom of the tube using a gel-loading tip and then transfer the aliquot to the center of the plastic dish.
5. Collect the embryos from the cold storage medium, transfer them to the washing dish, and wash them three times. Note: If you cannot retrieve all of the stored embryos, rinse the inside of the tube using 200 μL of M2 medium at the edge of the plastic dish.
6. Transfer the embryos into the oviducts of a pseudopregnant mouse.

Acknowledgments

The authors thank Yuka Horikoshi, Shiori Takeuji, Kiyoko Yamashita, Tomoko Kondo, Yukie Haruguchi, Yumi Takeshita, Yuko Nakamuta, and Shuuji Tsuchiyama (CARD, Kumamoto University) for technical support. This work was partially supported by grants of the National Bioresource Project from the Ministry of Education, Culture, Sports, Science and Technology of Japan and Research on Development of New Drugs from the Japan Agency for Medical Research and Development.

References

1. Takeo T, Tsutsumi A, Omaru T, Fukumoto K, Haruguchi Y, Kondo T, Nakamuta Y, Takeshita Y, Matsunaga H, Tsuchiyama S, Sakoh K, Nakao S, Yoshimoto H, Shimizu N, Nakagata N (2012) Establishment of a transport system for mouse epididymal sperm at refrigerated temperatures. Cryobiology 65 (3):163–168. https://doi.org/10.1016/j.cryobiol.2012.06.002
2. Yoshimoto H, Takeo T, Irie T, Nakagata N (2017) Fertility of cold-stored mouse sperm is recovered by promoting acrosome reaction and hyperactivation after cholesterol efflux by methyl-beta-cyclodextrin. Biology of Reproduction 96(2):446-455
3. Yoshimoto H, Takeo T, Nakagata N (2017) Dimethyl sulfoxide and quercetin prolong the survival, motility, and fertility of cold-stored mouse sperm for 10 days†. Biology of Reproduction 97(6):883-891
4. Takeo T, Hoshii T, Kondo Y, Toyodome H, Arima H, Yamamura K, Irie T, Nakagata N (2008) Methyl-beta-cyclodextrin improves fertilizing ability of C57BL/6 mouse sperm after freezing and thawing by facilitating cholesterol efflux from the cells. Biol Reprod 78 (3):546–551. https://doi.org/10.1095/biolreprod.107.065359
5. Takeo T, Nakagata N (2011) Reduced glutathione enhances fertility of frozen/thawed C57BL/6 mouse sperm after exposure to methyl-beta-cyclodextrin. Biol Reprod 85 (5):1066–1072. https://doi.org/10.1095/biolreprod.111.092536
6. Nakagata N, Takeo T, Fukumoto K, Kondo T, Haruguchi Y, Takeshita Y, Nakamuta Y, Matsunaga H, Tsuchiyama S, Ishizuka Y, Araki K (2013) Applications of cryopreserved unfertilized mouse oocytes for in vitro fertilization. Cryobiology 67(2):188–192. https://doi.org/10.1016/j.cryobiol.2013.06.011
7. Takeo T, Fukumoto K, Kondo T, Haruguchi Y, Takeshita Y, Nakamuta Y, Tsuchiyama S, Yoshimoto H, Shimizu N, Li MW, Kinchen K, Vallelunga J, Lloyd KC, Nakagata N (2014) Investigations of motility and fertilization potential in thawed cryopreserved mouse sperm from cold-stored epididymides. Cryobiology 68(1):12–17. https://doi.org/10.1016/j.cryobiol.2013.10.007
8. Takeo T, Nakagata N (2010) Combination medium of cryoprotective agents containing L-glutamine and methyl-{beta}-cyclodextrin in a preincubation medium yields a high fertilization rate for cryopreserved C57BL/6J mouse sperm. Lab Anim 44(2):132–137. https://doi.org/10.1258/la.2009.009074
9. Takeo T, Kondo T, Haruguchi Y, Fukumoto K, Nakagawa Y, Takeshita Y, Nakamuta Y, Tsuchiyama S, Shimizu N, Hasegawa T, Goto M, Miyachi H, Anzai M, Fujikawa R, Nomaru K, Kaneko T, Itagaki Y, Nakagata N (2010) Short-term storage and transport at cold temperatures of 2-cell mouse embryos produced by cryopreserved sperm. J Am Assoc Lab Anim Sci 49(4):415–419
10. Horikoshi Y, Takeo T, Nakagata N (2016) N-acetyl cysteine prolonged the developmental ability of mouse two-cell embryos against oxidative stress at refrigerated temperatures. Cryobiology 72(3):198–204. https://doi.org/10.1016/j.cryobiol.2016.05.002

Chapter 18

Reprogramming of Primary Human Cells to Induced Pluripotent Stem Cells Using Sendai Virus

Julia M. Draper and Jay L. Vivian

Abstract

Induced pluripotent stem (iPS) cells are important tools for studying differentiation and for use in patient-specific disease modeling. We present a detailed method for the reprogramming of primary human fibroblasts to induced pluripotent stem cells using Sendai virus. These procedures allow for the efficient generation of multiple high-quality feeder-independent iPS cell lines for a given human fibroblast line. The iPS cell lines generated by this protocol can be used in a variety of differentiation and gene expression studies, as well as in genetic manipulations.

Key words Human pluripotent stem cells, Induced pluripotent stem cells, Reprogramming, Primary cell, Sendai virus, Fibroblast

1 Introduction

Pluripotent stem cells are powerful research tools for understanding differentiation and gene regulation, and are increasingly being considered for therapeutic use [1]. Induced pluripotent stem (iPS) cells are somatic cells which have been reprogrammed to a pluripotent state that closely resembles embryonic stem cell phenotypes [2]. iPS cells are generated via expression of a cocktail of transcriptional regulators which reprogram the somatic cells to a pluripotent phenotype. Patient-specific iPS cells have the potential for use in studying many diseases of genetic and perhaps epigenetic origin, in that potential disease-causing variants or genetic alterations are captured in a cell type amenable for laboratory studies [3]. The pluripotent nature of iPS cells allows for their differentiation to virtually any cell type, including those most relevant for a particular disease study.

A variety of methods have been suggested for reprogramming human pluripotent stem cells, including different cocktails of transcription factors and/or small molecule modulators of specific signaling pathways. Expression of the cocktail of factors first

Melissa A. Larson (ed.), *Transgenic Mouse: Methods and Protocols*, Methods in Molecular Biology, vol. 2066,
https://doi.org/10.1007/978-1-4939-9837-1_18,

described by Yamanaka [2] (OCT4, MYC, KLF4, SOX2) remain the most utilized method for reprogramming. Several different methods of introduction of these factors into somatic cells have been successfully used, including viral infection or transfections with plasmid, RNA, or protein [4–8].

Several key considerations are taken into account when determining what method of reprogramming is to be used. Ideally, the method of introduction of the reprogramming factors should be efficient, broadly applicable to many somatic cell types, require limited hands-on manipulation, and be cost-effective. The reprogramming method should also not interfere with later manipulations of the iPS cells once established, such as further genetic manipulation and differentiation. Once iPS cells are established, the cells no longer require the reprogramming vector, and in fact, their continued expression can interfere with differentiation. The use of viral infection to introduce the cocktail of reprogramming factors is exceptionally efficient and can be broadly applied to many somatic cell types. Viral-based expression of reprogramming factors often requires only a single application to the somatic cells. In contrast, transfection of protein or nucleic acid often requires multiple, perhaps daily, transfection of the primary cells to achieve complete reprogramming. However, integrating viral expression vectors such as retrovirus or lentivirus have the potential for sustained expression of the reprogramming components. It has been fortuitous that these lentiviral transgenes are often, but not always, silenced upon differentiation. Thus, methods to either remove the reprogramming virus or the use of nonintegrating virus can obviate these concerns.

Sendai virus, which is a nonintegrating RNA virus, has recently been used for reprogramming somatic cells to generate iPS cell lines [9, 10]. Sendai virus infection is highly effective in many cell types, and the viral genome tends to be lost in the highly proliferative reprogrammed iPS cells after continued culture. The commercial availability of Sendai virus has allowed many researchers access to this tool for reprogramming.

In this chapter, we provide a detailed procedure we have developed for isolation of primary fibroblasts from patients, their use in reprogramming to generate patient-specific iPS cell lines using Sendai virus, and the transition of these cells to feeder-free culture conditions. This cost-effective and efficient protocol results in high-quality validated iPS cell lines suitable for many procedures. The entire reprogramming timeline from primary fibroblasts to validated iPS cells takes 4 months. However, early passages of iPS clones can be cryopreserved after only 2 months.

2 Materials

2.1 Equipment

1. Refrigerator.
2. −20 °C freezer.
3. −80 °C freezer.
4. Liquid nitrogen storage tank.
5. Autoclave.
6. Laminar flow biological safety cabinet.
7. 5% CO_2, 37 °C, humidified incubator.
8. Centrifuge with swinging bucket rotor for 15-ml conical tubes.
9. Inverted cell culture microscope, with 4× and 10× objectives and phase contrast optics.
10. Aspirator (vacuum pump).
11. 37 °C water bath.
12. Pipetting aid for pipettes.
13. Micropipettors for 200 μl and 1000 μl tips.
14. Sterile surgical instruments: scissors, forceps.
15. Bright-Line™ Hemacytometer.
16. Tally counter for counting cells.
17. Mr. Frosty Freezing Container.
18. Ice bucket.

2.2 Sterile Disposable Cell Culture Plastics

All plasticware should be nonpyrogenic and sterile.

1. Individually wrapped 5-, 10-, and 25-ml pipettes.
2. Samco™ general-purpose transfer pipettes.
3. Pipette tips with barrier.
4. 15-ml and 50-ml polypropylene conical centrifuge tubes.
5. 1.8-ml cryovials, internally threaded.
6. 0.2 μm pore size filter sterilization bottles, PES membrane, 150 ml, 250 ml, and 500 ml sizes.
7. Tissue culture plates and dishes: 6-well, 24-well, and 100 mm.
8. Petri dishes for cutting up biopsies.

2.3 Cell Culture

1. Human Fibroblast Medium: 1× Knockout Dulbecco's Modified Eagle's Medium (DMEM), 10% Embryonic Stem Cell-qualified Fetal Bovine Serum (FBS), 2 mM L-Glutamine, 1% Non-Essential Amino Acids (NEAA), 55 μM 2-Mercaptoethanol, 100 U-μg/ml Penicillin–Streptomycin, 1 mg/ml Amphotericin B, 50 μg/ml Gentamicin, and 1× Plasmocin Prophylactic.

2. Human ES Medium: 1× Knockout Dulbecco's Modified Eagle Medium/Nutrient Mixture F-12 (DMEM/F-12), 20% Knockout Serum Replacement, 2 mM L-Glutamine, 1% Non-Essential Amino Acids (NEAA), 50 μM 2-Mercaptoethanol, 10 ng/ml human recombinant bovine Fibroblast Growth Factor (bFGF), and 100 U-μg/ml Penicillin–Streptomycin.
3. Mouse Fibroblast Medium: 1× KnockOut DMEM, 10% Fetal Bovine Serum (FBS), 2 mM L-Glutamine, 1% NEAA, 50 μM 2-Mercaptoethanol, 100 U-μg/ml Penicillin–Streptomycin, and 1× Plasmocin Prophylactic.
4. 2× Freezing Medium: 1× Knockout DMEM, 20% FBS, and 20% Dimethyl sulfoxide (DMSO).
5. Human ES Freezing Medium: 1× Human ES Medium, 40% Embryonic Stem Cell-Qualified FBS, and 10% DMSO.
6. Attenuated Trypsin Solution: 1× Phosphate-buffered Saline (PBS), 0.05% Trypsin, 1% Chicken Serum, and 500 μM Ethylenediaminetetraacetic acid (EDTA).
7. Gelatin from porcine skin.
8. Matrigel, ES-qualified.
9. Mitomycin C.
10. Liberase.
11. ReLeSR™ (STEMCELL Technologies).
12. mTeSR™ (STEMCELL Technologies).
13. Phosphate-buffered Saline (PBS), pH 7.4, no calcium, and no magnesium.
14. Hank's Buffered Salt Solution (HBSS).
15. Dispase II.
16. TrypLE Express Enzyme.
17. CytoTune®-iPS 2.0 Sendai Reprogramming Kit.
18. ROCKi (Rho kinase inhibitor; Y-27632).
19. mFreSR™ (STEMCELL Technologies).
20. Trypan Blue Solution.

3 Methods

3.1 Preparation of Gelatin-Coated Tissue Culture Dishes

1. Prepare a solution of 0.1% gelatin in water. For example, measure 300 mg of powdered gelatin and add to 300 ml water.
2. Heat solution to dissolve gelatin via microwave or autoclave.
3. Filter solution with a 0.2 μm filter and store at room temperature.

4. To prepare gelatin-coated plates, coat tissue culture-treated plates with gelatin solution and incubate at room temperature for 40 min.
5. Aspirate gelatin solution and allow to dry overnight. Plate is ready to use. Plates can be stored at room temperature for up to 6 weeks.

3.2 Preparation of Matrigel-Coated Plates

1. Incubate tissue-culture plates and serological pipettes at 2–8 °C for at least 20 min.
2. Aliquot 24 ml DMEM/F-12 and put on ice for at least 20 min.
3. Put a 2 mg aliquot of Matrigel on ice.
4. Once plates and medium are cold, mix Matrigel with ice-cold DMEM/F-12. If Matrigel is still frozen, thaw by adding ice-cold DMEM/F-12 and pipetting to mix (*see* **Note 1**).
5. Remove plates from refrigerator and coat with Matrigel solution:

   6 ml per 100 mm dish.

   1 ml per well of 6-well plate.

   0.5 ml per well of a 24-well plate.
6. Incubate coated plates at room-temperature on a level surface for 1 h.
7. Plates can be wrapped with Parafilm and stored at 2–8 °C for up to 2 weeks. Aspirate Matrigel solution immediately before use.

3.3 Thawing Cells

1. Add 6 ml of prewarmed medium to a 15 ml conical tube.
2. Remove a vial of cells from storage and submerge the bottom half of the tube in a 37 °C water bath. Do not submerge the cap in water.
3. When only a small chunk of ice remains in the vial, remove from the heated bath and wipe with 70% ethanol.
4. Add 1 ml prewarmed medium from the 15-ml conical tube to the vial of cells dropwise using a P1000 pipette, stirring gently with the tip. Adding medium slowly (over the course of 20 s) reduces the risk of osmotic shock.
5. Gently transfer contents of the vial to a 15 ml conical tube.
6. Centrifuge cells in the conical tube for 3 min at 1000 × *g*.
7. Aspirate supernatant and resuspend the cell pellet by flicking the tube.
8. Gently resuspend the pellet in medium and transfer to a plate.

Table 1
Volumes for plate coating, feeding, and passaging

Plate size	Amount of gelatin for coating	Amount of medium for feeding	Volume of trypsin for passaging
100 mm	5 ml	9–12 ml	3 ml
6 well plate (one well)	1 ml	2–3 ml	1 ml
24 well plate (one well)	0.5 ml	0.5–1 ml	0.5 ml

3.4 Passaging Mouse and Human Fibroblasts with Trypsin or TrypLE

1. Aspirate growth medium and rinse well with PBS (no calcium, no magnesium). Aspirate the PBS rinse.
2. Add sufficient Trypsin or TrypLE to cover the cells (Table 1), and incubate for 5–10 min at 37 °C, or until cells detach from the plate and any clumps become translucent and loosely adherent.
3. Add a volume of 10% serum-containing medium equivalent to the trypsin volume to inhibit the reaction. Triturate cell suspension 5–15 times by gentle pipetting until a single-cell suspension is achieved, as monitored under a microscope.
4. Transfer the cell suspension to a 15-ml conical tube and centrifuge for 3 min at 1000 × *g*.
5. Aspirate supernatant and flick the tube to resuspend the pellet.
6. Resuspend the pellet in medium and plate cells onto gelatin-coated tissue-culture plates (*see* Subheading 3.1) at desired density. For early passage mitotically-active fibroblasts, a 1:5 split is recommended.

3.5 Preparation of Mouse and Human Fibroblasts for Cryopreservation

1. Follow Subheading 3.4 through **step 5**.
2. If counting cells prior to cryopreservation, resuspend pellet in a small amount of medium (2–3 ml for one well of a 6-well plate) and refer to Subheading 3.7 for further instructions.
3. Flick tube to resuspend the cell pellet in residual medium and add 0.5 ml of regular growth medium per vial of cells to be frozen. For example, to freeze cells into four vials, add 2 ml of regular growth medium.
4. Slowly add an equal volume of ice-cold 2× Freezing Medium and mix by gentle inversion.
5. Transfer cell suspension to cryovials.
6. Place cryovials in a Mr. Frosty freezing container and incubate at −80 °C for 24 h.

7. Fibroblasts can remain at −80 °C for several months. For long-term storage, transfer cryovials to liquid nitrogen.

3.6 Preparation of Inactivated Mouse Embryonic Fibroblasts (MEFs)

1. Thaw active MEFs and plate onto a gelatin-coated tissue culture dish (Subheading 3.1) in Mouse Fibroblast Medium.
2. Feed every other day until confluent.
3. Split cells using trypsin (Subheading 3.4) at a 1:5 split onto gelatin-coated tissue culture dishes.
4. Feed with Mouse Fibroblast Medium 24 h later, and again at 72 h.
5. After 4 days, mitotically inactivate cells using mitomycin C (*see* **Note 2**).
   (a) Prepare a solution of 10 μg/ml mitomycin C in sufficient Mouse Fibroblast Medium to cover cells. For example, 5 ml for a 100 mm dish.
   (b) Incubate for 3 h at 37 °C.
   (c) Aspirate medium and rinse plate with PBS.
   (d) Feed with Fibroblast Medium and allow cells to recover for 24 h.
6. Freeze cells into vials of 1.5×10^6 cells per vial. Thaw one vial to one 100 mm dish or 6-well plate to create a feeder layer for iPS cell culture (*see* **Note 3**).

3.7 Quantification of Cell Number with Bright-Line™ Hemocytometer

1. Follow passaging protocol (Subheading 3.4) through **step 5**.
2. Resuspend the cell pellet in medium. Use 2–3 ml of medium for one well of a 6-well plate. If cells are too sparse, centrifuge again and resuspend in a smaller amount of medium.
3. Remove 25 μl of cell suspension and combine with 25 μl of trypan blue solution. Flick the tube to mix.
4. Place a glass coverslip over the hemocytometer.
5. Add about 10 μl of blue cell suspension to the counting chamber, allowing liquid to enter under the coverslip via capillary action.
6. Count four squares at 10× magnification.
7. Calculate cells per ml as follows: average number of cells per square × 2 (dilution factor) × 10,000 = number of cells per ml.

3.8 Preparation of Liberase™ Solution for Dissociation of Biopsy Tissue

1. Rehydrate one 5 mg bottle of Liberase™ in 2 ml of ice-cold Water For Injection (WFI).
2. Bring volume up to 19 ml in solution of ice-cold KnockOut DMEM with 1 μg/ml Amphotericin B, 50 μg/ml Gentamicin, 100 U/ml Pen Strep, and Plasmocin Prophylactic. Solution should be 1.4 Wünsch units per ml.
3. Aliquot and store at −20 °C.

3.9 Passaging of Feeder-Dependent iPS Cells Using Dispase

1. Prepare a plate of mitotically inactive MEFs 24–48 h before passaging iPS cells (Subheading 3.6). Optional: prepare a gelatin-coated plate for feeder-depletion if cells to be passaged are growing on a dense feeder layer.
2. Aspirate medium from well to be passaged.
3. Rinse well with prewarmed DMEM/F-12; aspirate.
4. Add sufficient room temperature 5 units/ml Dispase (in DMEM/F-12) to cover well (Table 1).
5. Incubate cells in Dispase solution in 37 °C incubator for 5–15 min, or until edges of colonies begin to peel away from feeder layer.
6. Aspirate Dispase solution, and gently rinse well three times with DMEM/F-12.
7. Add Human ES Medium to the well and triturate until colonies have broken up into smaller chunks.
8. Optional: if feeder layer from the original plate is too dense, transfer cell suspension to a gelatin-coated plate and place in 37 °C incubator for 45–60 min. The MEFs will adhere to the plate, and the colonies can be rinsed off and plated onto a new feeder layer.
9. Plate at a 1:4 to 1:12 split. Cells can be passaged again in 5–7 days.

3.10 Passaging of Feeder-Independent Cells Using ReLeSR

1. Cells are ready to passage when colonies are large and compact with dense centers, and only just beginning to touch (*see* **Note 4**).
2. Prepare a Matrigel-coated plate (Subheading 3.2).
3. Rinse well to be passaged with PBS; aspirate.
4. Rinse well with ReLeSR (Table 1); aspirate within 60 s.
5. Incubate "dry" plate at room temperature for 5 min.
6. Add 6 ml of room temperature mTeSR (*see* **Note 5**). Undifferentiated colonies should pop up off the surface of the plate with gentle rocking.
7. Using a 6-ml serological pipette, transfer the cell suspension to a 15-ml conical tube. Triturate gently 3–5 times, or until cell aggregates are 50–200 um in diameter.
8. Aspirate Matrigel solution from a new plate, and directly plate clumps at a 1:16 split. Do not centrifuge cell suspension prior to plating.
9. Colonies will be ready to split again in 4–6 days (*see* **Note 6**).

3.11 Dissociation of Patient Biopsy Samples for Generation of Primary Fibroblasts (See Note 7)

All procedures using patient tissues must be approved by the Institutional Review Board along with associated patient approval forms. Tissues and cells derived from human patient samples should be manipulated and cultured with care under Biosafety Level 2 (BSL-2) conditions.

1. Place recently isolated live biopsy in tube of room temperature Roswell Park Memorial Institute (RPMI) or DMEM for transport.
2. Place biopsy onto sterile petri dish and cut into small pieces (<4 mm) using sterile surgical scissors and tweezers.
3. Transfer biopsy tissues and medium to a 15-ml conical tube ("Tube 1").
4. Centrifuge for 3 min at 1000 × *g*; discard supernatant in biohazardous waste container.
5. Resuspend pellet in 1–2 ml Liberase™ solution (Subheading 3.8) and incubate 30 min at 37 °C.
6. Without centrifuging, transfer supernatant to a new 15-ml conical tube ("Tube 2"). Rinse clumps of tissue once with sterile HBSS, and transfer to Tube 2 as well.
7. Add 1–2 ml additional Liberase™ solution to the remaining tissue in Tube 1 and incubate an additional 30 min.
8. Centrifuge Tube 2 (containing supernatant from first Liberase incubation) for 3 min at 1000 × *g*. Discard supernatant in biohazardous waste container.
9. Rinse pellet with sterile HBSS to remove residual Liberase and centrifuge again, discarding supernatant in biohazardous waste container.
10. Resuspend pellet in 9 ml of Human Fibroblast Medium and transfer to one well of a gelatin-coated 6-well tissue culture plate (Subheading 3.1). Grow cells at 37 °C, 5% CO_2, and 100% humidity.
11. Prepare pieces of tissue from Tube 1 in the same manner as the supernatant from Tube 2. Empty pieces onto three wells of a gelatin-coated 6-well tissue culture plate into 3 ml of medium per well.

3.12 Feeding and Cleaning the Dissociated Biopsy

1. If the biopsy was oily (e.g., from fatty tissues), aspirate spent medium from the top of the liquid, trying to suck up as much floating lipid as possible, then add more medium. Do not allow floating lipid layer to come in contact with cells, leaving some medium on the dish after daily feeding until the oil is eventually eliminated.

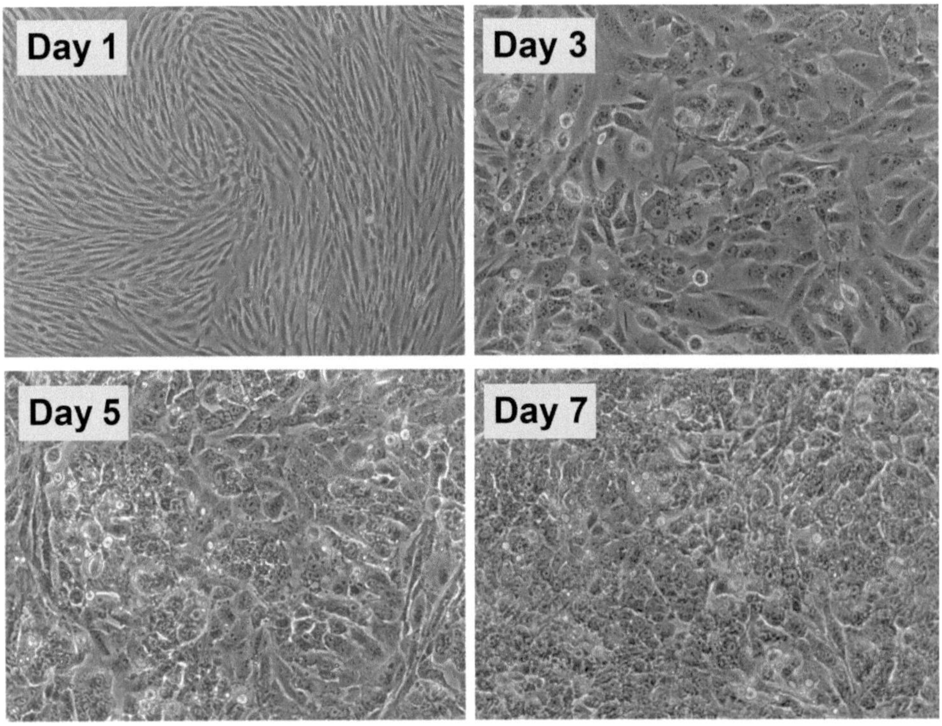

Fig. 1 Sendai virus reprogramming of primary adherent fibroblasts. Images of cells (10×) after viral infection 24 h, 72 h, 5 days, and 7 days

2. If biopsy culture does not have extensive floating lipid, aspirate medium and rinse with PBS to remove dead cells and other particulates before feeding, then add fresh medium.
3. Feed daily with Human Fibroblast Medium until plate is clean of nonadherent debris, then feed every other day.

3.13 Passaging and Freezing Primary Fibroblasts

1. After about 10 days in culture, you should be able to find areas of dense fibroblast outgrowth from single cells or small adherent clumps of tissue. At this point the cells can be passaged with TrypLE or Attenuated Trypsin Solution.
2. Pool cells from three to four fibroblast outgrowths into one well of a gelatin-coated 6-well plate (Subheading 3.1), and feed every other day with Human Fibroblast Medium for about 5 days or until confluent (Fig. 1, Day 1).
3. Dissociate cells using TrypLE or Attenuated Trypsin Solution, and freeze (Subheading 3.5) into two vials for each well of a 6-well plate (*see* **Note 8**).

3.14 Sendai Infection of Primary Human Fibroblasts

1. Day 1: Thaw about 500,000 primary human fibroblasts onto each of two wells of a gelatin-coated 6-well tissue-culture plate (Subheading 3.1) in Human Fibroblast Medium.

Table 2
Sendai virus quantities

Plate type	Number of cells on day of transduction	Volume of fibroblast medium for transduction	Titer (cell infectious units) per well of hKOS	Titer (cell infectious units) per well of hc-Myc	Titer (cell infectious units) per well of hKlf4
24-well	60,000	200–500 μl	3×10^5	3×10^5	2×10^5
6-well	300,000	1 ml	1.5×10^6	1.5×10^6	1.0×10^6

2. Day 2: Aspirate spent medium and replace with fresh Human Fibroblast Medium.
3. Day 4: Passage cells with TrypLE (Subheading 3.4) onto another gelatin-coated plate at a density of 16,000 cells per cm^2.
   (a) 30,000 cells per well of a 24-well plate.
   (b) 150,000 cells per well of a 6-well plate.
4. On day 6, harvest one well via TrypLE and use it to estimate cell number.
   (a) Goal of 60,000 cells per well on a 24-well plate.
   (b) Goal of 300,000 cells per well on a 6-well plate.
5. Add Sendai Virus to an aliquot of Human Fibroblast Medium (Table 2), then aspirate spent medium from well and replace with virus-containing medium.
6. Days 7–11: Feed daily with Human Fibroblast Medium. Cell morphology will change (Fig. 1).

3.15 Transition from Fibroblast to Stem Cell Phenotype

1. On day 11, plate mitotically inactive MEFs (Subheading 3.6) at a low density (Fig. 2a) onto several gelatin-coated 100 mm tissue culture dishes (Subheading 3.1). Make four dishes for fibroblasts infected in a 24-well plate. Make ten dishes for fibroblasts infected in a 6-well plate.
2. Seven days after the Sendai infection, use TrypLE to dissociate infected fibroblasts and passage (Subheading 3.4) onto inactivated MEFs in Human Fibroblast Medium. Plate at various densities. For example, 50,000 vs. 100,000 cells per 100 mm dish.
3. The next day, aspirate spent Human Fibroblast Medium and feed with Human ES Medium.
4. Feed daily with Human ES Medium for about 30 days, or until stem-like colonies emerge and are large enough to pick (Fig. 2b).

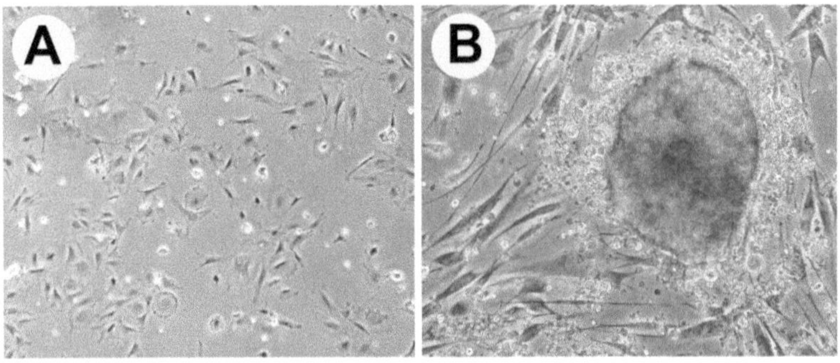

Fig. 2 Clonal expansion of iPS cell lines on MEF feeders. (**a**) Appropriate density of postmitotic mouse embryonic fibroblasts (4× magnification) 24 h postplating. (**b**) iPS cell clone plated onto MEFs and cultured for 20 days and ready for picking

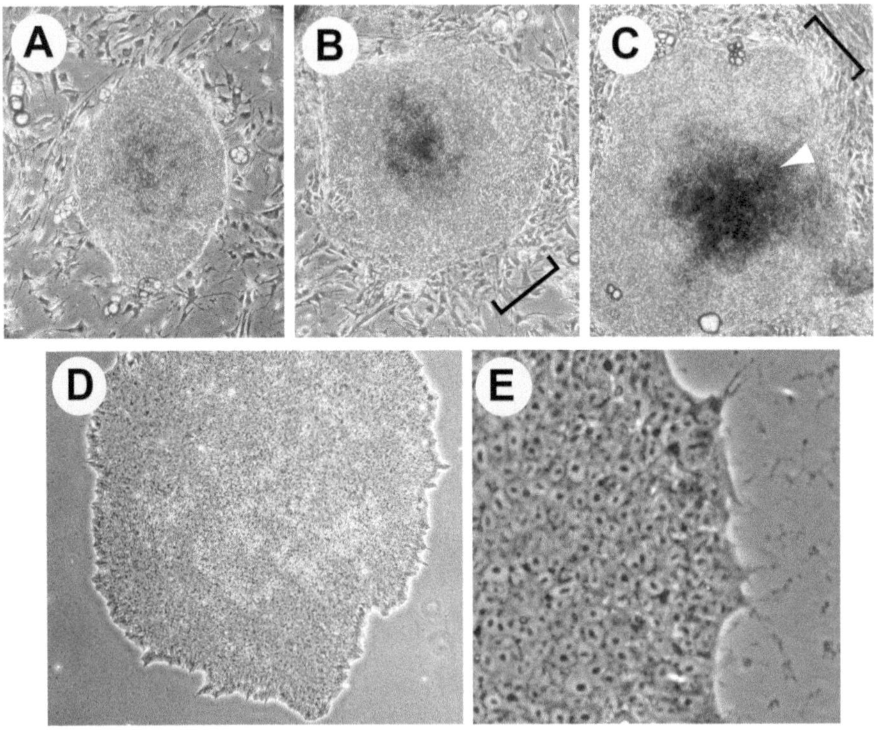

Fig. 3 Human iPS cell features as assessed via microscopy. (**a–c**) Primary iPS cell clones from Sendai virus reprogramming on MEFs with various morphologies. (**a**) Good-quality iPS cell clone with distinct edges. (**b**, **c**) Poorer quality clone with edge of colony with loosely coherent mesenchymal cells along edge of clone brackets. Note dark interior region (arrowhead, **c**) indicating thickened interior of clone, suggesting that this clone should have been picked and broken a few days prior. (**d**, **e**). iPS cells in extended culture on Matrigel. Note periphery of colony with clean and distinct edges. (**e**). Magnification of iPS cell clone to visualize cell morphology, noting large nuclei and distinct dark nucleoli

3.16 Picking Colonies

1. After 20–30 days growing in Human ES Medium, colonies should be large enough to pick (Fig. 2b).
2. Look for large colonies that have the following characteristics at higher magnification (*see* Fig. 3a–c):
   (a) Defined, round edges with no differentiation.
   (b) Uniform texture: small round cells with large nucleoli.
3. The day before picking colonies, prepare three gelatin-coated 24-well plates (Subheading 3.1) with inactivated MEFs (Subheading 3.6) at the appropriate density.
4. Immediately before picking colonies, aspirate Fibroblast Medium from 24-well plate and replace with 1 ml per well of Human ES Medium, supplemented with 10 μM ROCKi.
5. Pick 48–72 colonies using a P200 micropipettor:
   (a) Set micropipettor at 200 μl. Fully depress the pipette before submerging it into the medium.
   (b) Use the pipette tip to draw a circle around the colony, loosening it from the surrounding MEFs.
   (c) Draw several horizontal cuts through the colony using the edge of the pipette tip.
   (d) Use several vertical sweeps to scoop up pieces of the colony into the mouth of the pipette. Most of the pieces should stick to the pipette.
   (e) Collect broken clumps of colonies into pipette tip by sucking up medium at medium speed. This should collect the colony and most of its associated pieces.
   (f) Transfer colony into a well of the 24-well plate and pipette up and down three to four times to mix.
6. When all wells are full, transfer plate to incubator for culturing overnight.
7. Feed colonies daily with 0.5–1 ml Human ES Medium.

3.17 Passaging iPS Cells from 24-Well Plate to 6-Well Plate

1. After about 7 days of growing in the 24-well plate, iPS colonies should be large enough to survive passaging. Discard any clones with poor plating efficiency or substantial differentiation.
2. Prepare several gelatin-coated 6-well tissue culture plates (Subheading 3.1) with inactivated MEFs (Subheading 3.6) at least 24 h before passaging. Immediately before passaging, replace medium with 2.5 ml Human ES Medium supplemented with 10 μM ROCKi.
3. Rinse one well of the 24-well plate with DMEM/F-12, then add 200 μl of 5 units/ml Dispase (in DMEM/F-12).
4. Monitor colony detachment at room temperature initially, moving to 37 °C incubator if necessary. When edges of colonies

begin to peel up, carefully aspirate Dispase solution from plate and slowly add 200 μl DMEM/F-12, then aspirate. Repeat DMEM/F-12 wash two more times. Be careful not to dislodge colonies from plate while rinsing. They detach very easily at this stage.

5. If cell layer begins to dislodge, stop aspirating medium and move on to next step.
6. Add 500 μl of Human ES Medium supplemented with 10 μM ROCKi to the well (on top of the DMEM/F-12 if colonies have already detached) and triturate three times with a P1000.
7. Transfer contents to one well of a 6-well plate of adherent MEFs in Human ES Medium, and rock plate to evenly distribute clumps of cells.
8. Feed cells daily with Human ES medium.
9. Cells can now be passaged every 5–7 days. Discard clones with poor plating efficiency or substantial differentiation. Refer to Subheading 3.9 for passaging feeder-dependent iPS cells.

3.18 Freezing iPS Cells Growing on MEFs

1. After passage four, new iPS clones can be cryopreserved.
2. Begin passaging cells according to Subheading 3.9, **steps 2–7**. Be careful not to overtriturate colonies.
3. Transfer cells to a 15-ml conical tube, and centrifuge for 5 min at 300 × *g*. Aspirate supernatant.
4. Flick tube to resuspend cells in residual supernatant, then add ice-cold Human ES Freezing Medium. For cells from one well of a 6-well plate, 2 ml (two cryovials) is recommended.
5. Gently mix cells using a serological pipette and transfer to cryovials.
6. Place cryovials in a Mr. Frosty and place in a −80 °C freezer for 24 h.
7. After 24 h at −80 °C, cryovials containing iPS cells must be transferred to liquid nitrogen for long-term storage.
8. Thaw in medium supplemented with 10 μM ROCKi.

3.19 Transitioning iPS Cells from MEFs to Matrigel

Newly made iPS cells can be transitioned to feeder-free conditions after passage 4 (*see* **Note 9**). Select eight to ten clones with good morphology and a relatively fast growth rate to transition to feeder-free conditions. About half of them can be expected to fully transition while maintaining good morphology (*see* **Note 10**) (Fig. 3d, e).

RNA should be collected from cells after every 2–4 passages for use in analysis of the induced pluripotent stem cells. This RNA can also be used for quantifying the Sendai virus load in the iPS cells lines.

1. The first passage onto a Matrigel-coated plate should be done at a 1:1 split. This procedure is easiest if the cells are growing in a 100-mm dish.
2. For each 100-mm dish to be passaged, prepare one Matrigel-coated 100-mm dish (Subheading 3.2) and one gelatin-coated 100-mm dish (Subheading 3.1).
3. Rinse the 100-mm dish to be passaged with 4 ml of DMEM/F-12, and then add 3 ml of 5 units/ml Dispase (in DMEM/F-12).
4. Incubate at 37 °C, monitoring colony detachment every few minutes. When edges of colonies begin to peel up, aspirate Dispase solution and rinse three times with DMEM/F-12.
5. Add 6 ml of mTeSR supplemented with 10 μM ROCKi to the dish, and triturate three to four times with a serological pipette, or until colonies are sufficiently broken up.
6. Bring volume of cell suspension up to 10 ml in mTeSR supplemented with 10 μM ROCKi, and incubate in the gelatin-coated dish at 37 °C for 45–60 min. This allows some of the MEFs to adhere to the plate so that they do not persist after the transition to feeder-free culture.
7. Swirl plate to dislodge any iPS colonies that might be sitting on the bottom, and transfer to the Matrigel-coated plate. Feed daily for about 5–7 days, or until stem-like colonies are large enough to passage again.
8. For the second passage on Matrigel, rinse plate with 4 ml of PBS (no calcium, no magnesium), then rinse with 3 ml of ReLeSR. Aspirate ReLeSR, and incubate the "dry" plate at room temperature for 6–8 min.
9. Add 6 ml of mTeSR (without ROCKi) to the plate. At this point, it is fine to squirt medium very gently across the plate to dislodge colonies. However, most cells should remain on the plate.
10. Transfer colonies to a 15-ml conical tube, and triturate two to three times in the tube (*see* **Note 11**).
11. Plate onto a 100-mm Matrigel-coated tissue-culture dish at a 1:4 to 1:8 split.
12. Continue to passage with ReLeSR every 4–5 days.

3.20 Freezing Feeder-Independent iPS Cells

1. Passage cells with ReLeSR according to Subheading 3.10 through **step 7**.
2. Centrifuge cells in a 15 ml-conical tube for 5 min at 300 × *g*.
3. Aspirate supernatant and flick tube to resuspend cells.
4. Add ice-cold mFreSR and invert tube gently to mix.

5. Freeze cells at about twice the concentration that would be normal for a split. For example, a 100-mm dish of cells that would normally be split at 1:16 should be frozen into eight vials.
6. Place vials in a Mr. Frosty or similar device, and allow to freeze at −80 °C for 24 h.
7. After 24 h, transfer vials to liquid nitrogen.

Completion of this procedure will result in the generation of several clones of iPS cells for a given patient-specific fibroblast line. Further analysis of these cells will be important to determine which of these cell lines will be used for later analysis. RNA collected from these clones can be used to assess expression levels of pluripotency markers (OCT4, NANOG, SOX2, etc.) and Sendai virus load, either using RT-PCR or real-time PCR. iPS cells tend to lose the Sendai virus over extended passages, and thus clones with minimal or no detectable Sendai virus should be used for further experiments. Some Sendai virus kits have temperature-sensitive replication defective forms which allow for clearance of the virus via transient culture (~7 days) in slightly elevated temperature (38 °C), which does not otherwise impair the pluripotency. High-quality clones should also be analyzed for karyotype to confirm appropriate chromosome number, and this needs to be periodically monitored to confirm stable karyotype. The select clones that show quality morphology, lack of Sendai virus, and appropriate karyotype can then be used for experimental procedures, including use in a teratoma assay and differentiation.

3.21 Resources

WiCell Karyotyping: https://www.wicell.org/home/characterization/cytogenetics/karyotyping/g-banded-karyotyping.cmsx.

4 Notes

1. To prevent Matrigel from solidifying, do not warm above 8 °C.
2. MEFs can alternatively be irradiated for mitotic inactivation. Irradiators of different types (X-ray irradiator cabinet or gamma irradiation via ^{137}Cs source) have specific protocols for their use, and these procedures are beyond the bounds of this chapter. Mitomycin C is a convenient alternative for researchers who do not have access to such equipment.
3. A vial of frozen inactivated MEFs should be test-plated to confirm appropriate density prior to planning any iPS cell culture. Adjustments to plating density can be noted for future platings. This is particularly convenient if large numbers of MEFs are frozen in many cryovials in a single batch.

4. Avoid long culture times that result in differentiation, colonies growing together, or interior cells of the colony becoming differentiated or thickened and dark under phase optics.
5. mTeSR should be warmed to room temperature only. Exposure to 37 °C heat destabilizes the growth factors necessary for maintaining undifferentiated cultures.
6. Passaging procedure can be optimized as follows:
   (a) Start by incubating for 5 min after ReLeSR exposure, then adjust incubation time at the next passage depending on how well the colonies break apart in response to trituration.
   (b) Start by plating at a 1:16 split, and then adjust the split ratio at the next passage depending on how many colonies sat down.
   (c) If colonies take more than 6 days to regrow, triturate fewer times or at a slower speed.
7. Although this procedure has been optimized for dissociation of dermal fibroblasts, we have been successful in using this procedure in isolating primary fibroblasts from scar tissue from surgical procedures, such as removal of a feeding tube, and from brain dura from recently deceased autopsies.
8. The initial establishment of primary fibroblasts is performed in medium supplemented with a cocktail of antibiotics to ensure that cells are free from any common contamination. We have been successful in isolation of primary fibroblasts even from biopsies with potential for contamination, such as colon biopsies. Once the primary lines are frozen, further culture in these additional antibiotics beyond penicillin/streptomycin is not recommended.
9. At about passage 7, cells can be frozen from Matrigel culture with mFreSR and transitioned to a 6-well plate to save on media. Continue passaging cells on Matrigel for 15–20 passages to provide sufficient time for the cells to complete their reprogramming and establish a stable pluripotent phenotype.
10. Cells will exhibit poor-looking morphologies after the first passage onto Matrigel, and any stem-like colonies that appear will display a similar morphology to those grown on MEFs. After three passages on Matrigel, their morphology should change. Clones with sustained poor cell morphologies after the third passage on Matrigel should be discarded.
11. Colony size can be monitored under the microscope by transferring 50 μl of cell suspension to one well of a 96-well plate. When passaging colonies on Matrigel, cell clumps should be slightly smaller than the clumps that are produced from passaging iPS cell cultures on MEFs.

References

1. Keller G (2005) Embryonic stem differentiation: emergence of a new era in biology and medicine. Genes Dev 19:1129–1155
2. Takahashi K, Tanabe K, Ohnuki M et al (2007) Induction of pluripotent stem cells from adult human fibroblasts by defined factors. Cell 131:861–872
3. Wu SM, Hochedlinger K (2011) Harnessing the potential of induced pluripotent stem cells for regenerative medicine. Nat Cell Biol 13:497–505
4. Shao L, Wu WS (2010) Gene-delivery systems for iPS cell generation. Expert Opin Biol Ther 10:231–242
5. Warren L, Manos PD, Ahfeldt T et al (2010) Highly efficient reprogramming to pluripotency and directed differentiation of human cells with synthetic modified mRNA. Cell Stem Cell 7:618–630
6. Yu J, Hu K, Smuga-Otto K et al (2009) Human induced pluripotent stem cells free of vector and transgene sequences. Science 324:797–801
7. Zhou H, Wu S, Joo JY et al (2009) Generation of induced pluripotent stem cells using recombinant proteins. Cell Stem Cell 4:381–384
8. Zhou W, Freed CR (2009) Adenoviral gene delivery can reprogram human fibroblasts to induced pluripotent stem cells. Stem Cells 27:2667–2674
9. Ban H, Nishishita N, Fusaki N et al (2011) Efficient generation of transgene-free human induced pluripotent stem cells (iPSCs) by temperature-sensitive Sendai virus vectors. Proc Natl Acad Sci U S A 108:14234–14239
10. Fusaki N, Ban H, Nishiyama A et al (2009) Efficient induction of transgene-free human pluripotent stem cells using a vector based on Sendai virus, an RNA virus that does not integrate into the host genome. Proc Jpn Acad Ser B Phys Biol Sci 85:348–362

INDEX

A

Anesthesia 102, 105, 107, 108, 110, 206
Animal welfare 107

B

Behaviors 1, 172, 173, 176–182, 184, 186–188
Blastocysts 2, 4–7, 12–14, 48, 50, 54–56, 83–88, 102, 103, 108, 109, 120, 121, 135, 140–142, 164–166

C

2-cell stage 89–99
Cauda epididymis 129, 214
Chimeras 10, 12, 13, 60–62, 64, 67, 78, 79, 135, 138, 164–166, 168
Cloning 2, 8, 37, 39, 90, 92, 93, 113–122, 129
Clustered regularly interspaced short palindromic repeats (CRISPR) 14, 16, 27, 28, 43, 47–57, 60, 65, 93, 140–141, 143–145, 158, 167
Cognition 172, 178, 185–187
Cold storage 211–215
CRISPR associated protein (Cas9) 14, 16–18, 43, 48, 50, 53, 57, 60, 65, 70, 71, 74, 93, 95, 136, 138
CRISPR/Cas9 .... 13, 14, 16, 17, 43, 44, 47, 48, 50, 54, 134, 136–140, 142–143, 158, 163–168
crRNA 48, 50, 52, 53, 57

E

Electroporation 18, 48, 50–55, 57, 60, 62–65, 69–71, 89
Embryonic stem (ES) cells 2, 10–14, 16, 37, 59–81, 83–89, 93, 97, 98, 108, 114, 115, 117, 125, 134–136, 140–142, 156, 157, 160, 164, 165, 168, 217
Embryo transfers (ET) 5–7, 38, 50, 52, 97, 101–110, 116, 120–122, 195, 196, 202, 205–206, 211
Embryos 2–8, 12, 13, 18, 27–33, 37, 38, 40, 43, 47–57, 61, 79, 84–99, 101–110, 113, 114, 116, 120–122, 127–130, 136, 138, 157, 165, 166, 172, 196, 202–204, 207, 208, 211, 212

F

Fibroblasts 61, 62, 86, 117, 218–220, 222, 223, 225–229, 232, 233

G

Gene-editing 134, 158, 163, 172
Gene-modified .... v, 27, 28, 31, 83, 89, 101, 149–161
Generation of genetically engineered mice 17
Gene-targeting 10, 12, 16, 31, 59, 60, 62, 136, 163–168
Genetically engineered mice 1, 4, 133, 211–215
Genotyping 38, 40–42, 97, 133–147, 156, 164–166, 168
Germ line transmission 12, 61, 77, 136, 138, 157, 164–167
Guide RNA (gRNA) 14, 17, 47, 48, 52, 53, 57, 93, 136, 158, 164

H

H11 locus 37, 38, 41, 45
Homologous recombination 10, 12, 37, 43, 59, 60, 74, 80, 81, 89, 157, 165
Human pluripotent stem cells 217

I

In vitro fertilization (IVF) 4, 7, 101, 108, 127–130, 195, 201–202, 204, 205, 208, 211–215
In vivo 37, 47–57, 60, 116, 189
Indels 16, 43, 48, 74, 81, 89, 166
Induced pluripotent stem cells 217–233

K

Knock-in 16, 43, 44, 89–99, 134, 136, 140, 163, 164
Knockouts (KO) 12, 14, 16, 43, 44, 64, 89, 157, 159, 189, 219, 220, 223

Melissa A. Larson (ed.), *Transgenic Mouse: Methods and Protocols*, Methods in Molecular Biology, vol. 2066, https://doi.org/10.1007/978-1-4939-9837-1,

L

Locomotion 173, 181–182, 184, 186, 188

M

Microinjections 2, 3, 6–8, 10, 12–14, 16, 18, 27–33, 35–45, 50, 61, 64, 78–80, 83–93, 95–99, 125, 160
Micromanipulation 3, 4, 8
Mosaics 138, 164, 166, 167
Mouse 1, 27, 36, 47, 59, 83, 89, 101, 107, 114, 125, 134, 149, 163, 172, 195, 211, 220
  genetics 2, 89, 149, 150, 161
  strains 2, 5, 7, 48, 63, 74, 110, 115, 149, 150, 172, 187

N

Names 149–160, 207
NEPA21 .... 50, 53
Nomenclature 149–161
Non-surgical 6, 107–110, 122
Non-surgical embryo transfer device (NSET) 107–110
Nuclear transfer (NT) 113–122, 125

O

Oocytes 4, 7, 8, 96, 113–122, 126–128, 154, 158, 195, 196, 201, 202, 204, 205, 207, 208, 212, 214, 215
  activation 119, 120
  vitrification 204–205

P

Phenotypes 1, 2, 10, 16, 17, 151, 160, 164, 165, 172, 174–176, 178, 183, 185, 187, 188, 217, 233
PhiC31 integrase 36, 43
Polymerase chain reaction (PCR) 8, 16, 39–42, 48, 50, 52, 54–56, 60, 62, 65, 73–77, 81, 90, 93, 95, 99, 134–138, 140–146, 164, 167, 232
Primary cells 218
Pronuclear microinjection 27–33, 35–45, 96, 125, 160
Pseudopregnant2, 5, 6, 8, 18, 87, 95, 97, 101, 108, 109, 120, 196, 202, 205, 215

R

3Rs 107, 195
Recipients 2, 5, 6, 8, 32, 87, 95, 101–105, 107–110, 122, 128, 155, 160, 196, 205
Rederivation 108
Reproducible science 149
Reprogramming 114, 217–233
Rosa26 knock-in 37

S

Screening 10, 59, 73–77, 81, 97, 140, 142–145, 172, 173, 175, 177, 179, 182, 185, 187, 189
Sendai virus 217–233
Sensory functions 179–181
SHIRPA 175–177, 179
Site-specific transgenesis 36–38, 40–43
Somatic cells 113, 115, 217, 218
Spermatozoa 126, 127, 129, 205, 207, 208
Sperm cryopreservation 195, 198–200
Sperm-mediated gene transfer (SMGT) 126, 127
Standardized 4, 150, 151, 161, 172, 175, 179, 184, 189, 212
Superovulation 2–4, 7, 8, 48, 49, 51, 56, 84, 92, 95, 116, 128, 207
Surgery 101–106

T

TARGATT™ 36–40, 42–45
Targeted 10, 12, 16, 35–45, 57, 59–81, 83, 89, 134–141, 150, 156–160, 164–167, 189
Transgenes 8–13, 18, 36–39, 41–44, 90, 125–127, 130, 134, 135, 151, 157, 160–161, 164, 165, 218
Transgenesis 1–18, 36, 114, 125, 163, 164
Transgenics .... v, 2–4, 6–10, 13, 14, 18, 27, 28, 31, 35–45, 63, 83, 96, 125, 126, 134, 139, 154, 158, 160, 161, 163–165, 167, 171–189
  animals 13, 125
  core facilities 7, 12, 18, 28
  mice 2, 3, 8–10, 14, 27, 37, 38, 43, 134, 139, 160, 163, 165
Transportation 211–215
Two-cell embryos 4, 32, 129, 195, 196, 202, 205, 212–215
Two-cell embryo vitrification 202

V

Validation 47–57
Vasectomized males ...38, 101–103, 105, 108, 129, 196, 205

Z

Zygotes .2–8, 10, 14, 16, 18, 29, 30, 36, 37, 40, 48, 53, 54, 56, 89, 125

MIX
Papier aus verantwortungsvollen Quellen
Paper from responsible sources
FSC® C105338

If you have any concerns about our products,
you can contact us on
ProductSafety@springernature.com

In case Publisher is established outside the EU,
the EU authorized representative is:
Springer Nature Customer Service Center GmbH
Europaplatz 3, 69115 Heidelberg, Germany

Printed by Libri Plureos GmbH
in Hamburg, Germany